U0348952

Influence of grazing system on herbage production and animal performance in the Inner Mongolian steppe, China

Jun Hao

中国农业科学技术出版社

图书在版编目（CIP）数据

放牧系统对中国内蒙古草原牧草生长及动物生产性能的影响＝Influence of grazing system on herbageproduction and animal performance in the Inner Mongolian steppe，China：英文／郝俊（Jun Hao）著．—北京：中国农业科学技术出版社，2019.4

ISBN 978-7-5116-4081-9

Ⅰ.①放…　Ⅱ.①郝…　Ⅲ.①放牧管理-影响-草原生态系统-研究-中国-英文　Ⅳ.①S812.29

中国版本图书馆 CIP 数据核字（2019）第 050833 号

责任编辑　闫庆健　陶　莲
责任校对　贾海霞

出 版 者　中国农业科学技术出版社
北京市中关村南大街 12 号　邮编：100081
电　　话　（010）82109705（编辑室）　（010）82109704（发行部）
（010）82109709（读者服务部）
传　　真　（010）82106625
网　　址　http://www.castp.cn
经 销 者　各地新华书店
印 刷 者　北京建宏印刷有限公司
开　　本　710mm×1 000mm　1/16
印　　张　8.5
字　　数　228 千字
版　　次　2019 年 4 月第 1 版　2019 年 4 月第 1 次印刷
定　　价　88.00 元

Personal profile

Jun Hao, born in 1984, Shanxi, China. PhD. graduated from Kiel university, Germany. Associate Professor, chairman of the department of grassland science at Guizhou university. Member of the youth working committee of the Chinese Grassland Society. The research focuses on grassland grazing management and forage processing.

Preface

Livestock production is the main agricultural activity in Inner Mongolian steppe. Pastures nearby the farms are generally used for grazing throughout the whole grazing season. On the other hand, hay-making sites are often located farther away from farms and utilized for collecting hay for winter feeding.

However, the continuous grazing in the same areas might reduce vegetation cover, standing biomass, and also root biomass. Consequently, the risk of water and wind erosion significantly increased. More than 20% of the grassland areas in Inner Mongolia are not suitable for grazing and about 30% were degraded in the 1990s. Furthermore, the area of degraded grasslands in China has increased by about 2 million ha every year since 1990.

Grazing system is considered to be important for maintaining or even increasing the long-term grassland productivity and animal performance. The use of suitable grazing systems (e. g. rotational grazing) might mitigate the reported negative effects of heavy grazing. However, some research showed that the effects of rotational grazing on the quantity and quality of herbage and animal performance are inconsistent. Therefore, the limit factors to reveal the superiority of rotational grazing should be following with interest.

In this book, we aim to investigate the impacts of different grazing management systems and stocking rates on herbage species composition, herbage quality and quantity, digestibility of ingested herbage, feed intake, and live weight gain of sheep and to develop management concepts for a sustainable utilization of the grassland resources.

Jun Hao

March 2019

List of abbreviation

Abbreviation	Full name
ADF	Acid detergent fiber
ADL	Acid detergent lignin
ALT	Alternating
AN-PP	Above-ground net primary production
CON	Continuous
CP	Crude protein
CDOM	Cellulase digestible organic matter
DM	Dry matter
dOM	Digestibility of ingested organic matter
DO-MI	Digestible organic matter intake
FOM	Fecal organic matter
GI	Grazing intensity
GS	Grazing system
HA	Herbage allowance
HM	Herbage mass
LW	Live weight
LWG	Live weight gain
LWGs	Live weigh gain per sheep
LWGh	Live weight gain per hectare
M	Month
ME	Metabolizable energy

(attached table)

Abbreviation	Full name
MEI	Metabolizable energy intake
NDF	Neutral detergent fiber
NIRS	Near-infrared reflectance spectroscopy
OM	Organic matter
OMI	Organic matter intake
ROT	Rotational
SE	Standard error
SEM	Standard error of mean
SR	Stocking rate
TiO_2	Titanium dioxide
YE	Year

Contents

Chapter 1 General introduction

1 General introduction

1.1 Grasslands in Inner Mongolia

Grasslands in Inner Mongolia cover around 791 000 km^2 (Kawamura et al., 2005), and are thus one of the largest grassland ecosystems in China (approximately 20% of the total grasslands area; Liu et al., 2008). The Inner Mongolian steppe is under a semi-arid temperate climate, with annual rainfall of around 350 mm and dominated by drought tolerant plant species (Kang et al., 2007). The dominant soil type in this region is the chestnut soil (Chen and Wang, 2000). The steppe vegetation is dominated by 12 plant species. Among these species, *Leymus chinensis* Tzvel, *Stipa grandis* P. Smirn, *Festucasulcata* Beck, and *Artemisia frigid* Willd are the most common ones (Sun, 2005).

1.2 Grassland utilization in the Inner Mongolian steppe

Livestock production is the main agricultural activity in Inner Mongolian steppe. Farmers used the nomadic grazing systems several decades ago, which avoid continuous grazing in the same area due to the shift according to grassland conditions (Thwaites et al., 1998).

Human impacts are one of the main causes for the grassland degradation and desertification in Northern China in recent years (Yu et al., 2004). Sharply rising human populations increased the demands for the animal products and pressures on grassland ecosystems in the last 5 decades (Suttie et al., 2005). Since 1950, human population as well as the number of grazing cattle and in particular sheep in Inner Mongolia rapidly increased, which strongly reduced the ratio of available grassland per animal (Jiang et al., 2006). Yiruhan et al. (2001) reported that the available land area per sheep decreased from 6.8 ha in the 1950s to 1.6 ha in the 1980s in Inner Mongolia. For example, the human population in the Xilingol League, one of the biggest livestock areas inthe Inner Mongolia, increased from 200 000 in 1949 to approximately 950 000 in 2000 (Jiang et al., 2006). Correspondingly, the livestock numbers increased by 18-fold from 1949 to 2000, so that the available grassland area decreased from 5 ha to 1 ha/sheep (Jiang et al., 2006). Meanwhile, the current grazing system in the Inner Mongolian steppe consists of grasslands close to farms being utilized for intensive sheep and cattle grazing. Other, distant areas are only used for hay-making (Christensen et al., 2003). Farmers typically focus on the short-term economic output and neglect the long-term grassland productivity. Hence, they prefer to increase the stocking rate with the goal of maximizing profit in a short-term. As a result, the grazing pressure on the Inner Mongolian steppe has continuously increased, leading to severe ecological and economic problems (Jiang et al., 2006). Due to the increasing human population and the strong incentive by government for the nomadic families to settle and abandon their traditional utilization way of steppe, the grazing system in the Inner Mongolia steppe is characterized by spatial distinction between continuous grazed areas and hay-making sites (Christensen et al., 2003). Pastures nearby the farms are generally used for grazing throughout the whole grazing season. On the other hand, hay-making sites are often located farther away from farms and utilized for collecting hay for winter feeding (Schönbach et al., 2009). The continuous grazing in the same areas might reduce vegetation cover, standing biomass, and also root biomass. Consequently, the risk of water and wind erosion significantly increased (Lu et al., 2005; Zhao et al., 2005). Moreover, grassland in hay-making areas may lead to degradation due to the lack of nitrogen from animals' excretions. This has significant influences on the long-term grassland production (Owens and Shipitalo, 2009). The consequences of this grassland utilization change are the degradation and desertification of the Inner Mongolian steppe. More than 20% of the grassland areas in Inner Mongolia are not suitable for grazing and about 30% were degraded in the 1990s (Yu et al., 2004). Furthermore, the area of degraded grasslands in China has increased by about 2 million ha every year since 1990 (Li, 1999a). In 2004 about approximately 90% of the grasslands in China suffered to various degrees of degradation (Jiang et al., 2006). Tong et al. (2004) reported that the degradation and desertification of grasslands in northern China, particularly in Inner Mongolia, is the driving factor for the frequent sand and dust storms in China. Furthermore, the current grazing system in the Inner Mongolian steppe is far away from being of sustainable use. Future research efforts are necessary to combine the economic and ecological revenues of grasslands to evaluate the grazing management success (Yu et al., 2004).

Fortunately, the Chinese government attempted to protect the grasslands with land-use policies at the end of the 1990s. The policies focus on 4 aspects: ①seasonal grazing: dividing the grasslands into several areas by different seasons and grazing sequentially, based on the climate, species composition, water resources, and grazing management; ② grazing ban: forbidding grazing activities in the heavily destroyed grasslands; ③financial compensation: providing food and money refund to farmers in grazing ban areas; ④legal protection: establishing the "Grassland Law" for rational protection and utilization of grassland in 2002 (Chen and Wang, 2006; Zhang, 2008).

1.3 Effect of stocking rate on the grassland vegetation and animal production

Stocking rate is a critical factor affecting vegetation conditions and livestock production from the grasslands. A too low stocking rate results in waste herbage and low animal gain per unit area, while an over-stocking rate leads to degradation of grassland conditions and lower live weight gain per animal (Ralphs et al., 1990).

World - wide, many studies evaluated the vegetation responses to increasing stocking rates. The increasing grazing pressure changes its species composition on rough fescue grassland in Canada (Willms et al., 1985) and on a semi-arid typical steppe in Inner Mongolia, China (Zhao et al., 2007). Chen et al. (2003) and Wang and Li

(1999) indicated that the vegetation species in Xilin River Basin, Inner Mongolia are *Leymus chinensis* and *Stipagrandis* are replaced by *Cleistogenes squarrosa* under continuously heavy grazing, and finally replaced by *Artemisia frigid* Willd. These changes in the botanical composition of the vegetation may decrease the nutritional quality of the forage available for livestock grazing (Chen et al., 2003). Hence, the study by Chen et al. (2003) in Inner Mongolia indicated that crude protein content in forage samples decreased due to the species composition changes with increasing stocking rate by grazing sheep. Similarly, vegetation cover and standing biomass decreased with increasing stocking rate (Ralphs et al., 1990; Li et al., 2000). Moreover, Ferraro and Oesterheld (2002), Jiang et al. (2006), and Schönbach et al. (2011) reported that negative effects of grazing on above-ground net primary production of grasslands in different areas.

Due to the changes of vegetation caused by an increasing stocking rate, the digestibility of ingested herbage, intake of organic matter, and the live weight gain of animal would be affected. According to different chemical compositions of standing forage and greater capability of selection for higher quality herbage at low stocking rate, Garcia et al. (2003) speculated that digestibility of organic matter should be affected by stocking rates. The hypothesis was consistent with earlier study by Lu (1988), who reported that goats ingested herbage of a higher digestibility when the stocking rate is low. However, these results contradict the observations of Glindemann et al. (2009) and Müller et al. (2012), who conducted their studies in sheep in the Inner Mongolian steppe. This discrepancy may be explained by Schiborra (2007) who stated that diet selection was limited in this region due to a small difference in the digestibility of the herbage at different heights above ground level. Many studies reported that reduced organic matter intake is correlated with the increasing stocking rate (Hull et al., 1961; Langlands and Bennett, 1973). Nevertheless, some studies suggested that stocking rate did not affect animals' feed intake (Ackerman et al., 2001; Aharoni et al., 2004; Glindemann et al., 2009; Müller et al., 2012). Wang et al. (2005) concluded that animals prefer areas with a higher biomass to maintain their feed intake when grazing on a low herbage production grassland. Astocking rate largely influences live weight gain of animals and overall profitability of grazing management systems (Biondini et al., 1998). Quite a few studies estimated the effects of stocking rate on animal production with different species (cattle, sheep, and goat) and concluded that an increasing stocking rate may maximize live weight gain per unit of land in the short-term, however, reduced live weight gain per animal (Hull et al., 1961; Fynn and O'Connor, 2000; Han et al., 2000; Glindemann et al., 2009; Lin et al., 2012). Furthermore, Kemp and Michalk (2007) suggested that there might be a certain threshold, above which any increases in stocking rates do not contribute to or even reduce the total output per unit of land area due to the decreasing performance of individual animal. After 5 years of their study, Lin et al. (2012) concluded that in the Inner Mongolia steppe, this threshold was approximately 5.3 sheep/ha across the study years for the 3 months continuous grazing period, which ensures the long-term animal and grassland productivity as well as the ecosystem functions.

Grazing system is considered to be important for maintaining or even increasing the long-term grassland productivity and animal performance (Long, 1986). Therefore, the use of suitable grazing systems might mitigate the reported negative effects of heavy grazing. Allan (1997) compared rotational grazing, alternating grazing, and continuous grazing and concluded that rotational and alternating grazing were superior to continuous grazing with an increasing stocking rate. Moreover, Schönbach et al. (2011) reported that the above-ground net primary production was higher at alternating grazing system than at continuous grazing system in a moderate to heavy stocking rate after a 4 years study in the Inner Mongolian steppe. Similarly, Müller et al. (2012) evaluated the impacts of alternating grazing system and stocking rates on herbage quality and live weight gain of grazing sheep and suggested that alternating counteract the negative effects of heavy stocking rate on herbage and animal performance. However, they also mentioned that the positive effects of alternating system were limited by the low rainfall in the grazing season.

Table 1.1 Common grazing systems used in the livestock management studies.

Grazing systems	Definition and characteristic
Continuous grazing	grazing in a particular pasture the whole year or an entire grazing season (Hodgson, 1979)
Rotational grazing	animals are sequentiallygrazed from one pasture to another during an entire grazing season (Campbell, 1966)
Alternating grazing	grazing and hay-making areas are alternated between two or more than two pastures, the period of hay-making continuous for more than 12 consecutive months until next grazing period (Merrill, 1954)
Strip grazing	pasture is grazed in strips to enhance the utilization rate, giving livestock allocationto fresh pasture each day (Jamieson and Hodgson, 1979)
Rationed grazing	a predetermined amount of forage is allotted to the animal on a daily, weekly, or longer basis (Geus, 1950)
Extended grazing	grazing season is lengthened by utilization ofhay filed for pasture (D'souza et al., 1990)
Time limiting grazing	livestock graze a limited amount of high quality forage for a short period, usually once or twice daily, as a supplement to lower quality forage (Gerrish, 1993)

1.4 Grazing systems

World-wide, continuous heavy grazing has caused different degree of grasslands degradation (Gebremedhin et al., 2004). Managers attempt to find several methods to maintain and improve the grassland production, and thus enhance the animal performance. Specialized grazing systems (Table 1.1) with rest periods during the grazing season are considered to be effective to achieve the goals (Long, 1986). Periods with absence of animal grazing allow the vegetation to re-grow, to produce more seeds, and thus, to increase forage production and herbage nutrient quality (Heady, 1961). Li (1999b) suggested that efficient grazing management in the grassland ecosystems may protect the grassland from degradation and thereby maintain or even improve the long term grassland production and animal performance.

Many studies compared continuous (CON) grazing with specialized grazing system and attempted to obtain a more efficient grazing system to increase the grassland and livestock production. Several studies evaluated CON and rotational (ROT) grazing and reported that ROT increased herbage production (Virgona et al., 2000) and live weight gain (LWG) of individual animals (McMeekan and Walshe, 1963; Allan, 1997). On the contrary, Hafley (1996) and Derner et al. (2008) reported that lower LWG was observed at ROT than at CON. Furthermore, some studies stated that no differences were found in forage and animal parameters between ROT and CON by cattle (Manley et al., 1997; Popp et al., 1997) and sheep (Wang et al., 2009) grazing. Hence, Briske et al. (2008) reviewed the previous studies and concluded that ROT was not superior to CON. Similarly, inconsistent results have been also reported in the comparison of CON and alternating (ALT) system. The forage production and litter accumulation were observed to be higher at ALT than at CON (Clarke et at., 1943; Reardon and Merrill, 1976), and the above-ground net primary production of the grassland in Inner Mongolian steppe increased comparing ALT to CON at stocking rate from 6.0 to 9.0 sheep/ha (Schönbach et al., 2011). However, studies reported that ALT did not affect herbage nutrient quality (Schönbach et al., 2009) and productivity (Allan, 1985; Grant et al., 1988), and hence, live weight gain of individual sheep (Allan, 1997; Lin et al., 2012). As mentioned above, the effects of specialized grazing systems on herbage and animal parameters as well as grassland ecology were inconsistent in the previous studies. Results indicate that the specialized systems are not always better than CON. Positive effects of specialized grazing systems might be dependent on animal species used (Kitessa and Nicol, 2001), disparity of stocking rate applied (Warner and Sharrow, 1984), and differences in climates and ecosystems in the study areas (Van Poollen and Lacey, 1979).

1.5 Research project

This study is part of a long-term research project that evaluates the multiple effects of grazing on the typical steppe of Inner Mongolia. The goal is to elucidate the interaction between grazing of steppe ecosystems and matter fluxes, and to develop concepts for sustainable grassland utilization. The Sino-German research collaboration "Matter fluxes of Grasslands in Inner Mongolia as influenced by stocking rate" was funded by the German Research Foundation (No.536). The 11th sub-project contributed to the overall aim of the research group on site as well as on a regional scale:

(1) Amount, composition, and turnover of organic matter pools in grassland soils under typical steppe vegetation types of the Xilin River Basin as influenced by different grazing intensities. (German partner: Department of Soil Science, Technical University of Munich; Chinese partner: Department of Soil and Water Science, China Agricultural University, Beijing, and Institute of Botany, Chinese Academy of Science, Beijing).

(2) Effects of grazing intensity on net primary production and nutrient dynamics. (German partner: Institute of Plant Nutrition and Soil Science, Christian-Albrechts-Universitätzu Kiel; Chinese partner: Department of Plant Nutrition, China Agricultural University, Beijing).

(3) Impact of grazing management on yield performance, herbage quality, and persistence of grassland ecosystems of Inner Mongolia. (German partner: Institute of Grass and Forage Science/Organic Agriculture, Christian - Albrechts - Universitätzu Kiel; Chinese partner: Institute of Botany, Chinese Academy of Sciences, Beijing).

(4) Impact of grazing intensity on herbage quality, feed intake, and animal performance of grazing sheep in the grassland steppe of Inner Mongolia. (German partner: Institute of Animal Nutrition and Physiology, Christian-Albrechts-Universitätzu Kiel; Chinese partner: College of Animal Science and Technology, China Agricultural University, Beijing)

(5) Quantification and biogeochemical modeling of C and N turnover processes and biosphere - atmosphere exchange of C and N compounds. (German partner: Institute for Meteorology and Climate Research, KarlsruheInstitute ofTechnology; Chinese partner: Institute for Atmospheric Physics, Chinese Academy of Sciences, Beijing).

(6) Quantification of water and carbon exchange by micrometeorology and remote

sensing in managed steppe ecosystems of Inner Mongolia. (German partner: Institute for Hydrology and Meteorology, Technical University Dresden; Chinese partner: Institute for Atmospheric Physics, Chinese Academy of Sciences, Beijing).

(7) Regional water fluxes and coupled C and N transport. (German partner: Institute for Landscape Ecology and Resources Management, Justus-Liebig-Universität Gießen).

(8) Influence of various grazing intensities on soil stability and water balance on the plot scale (German partner: Institute for Plant Nutrition and Soil Science, Christian-Albrechts-Universitätzu Kiel; Chinese partner: Institute of Soil Science, Chinese Academy of Sciences, Beijing).

(9) Dynamics of wind erosion in the Xilin River Catchment area in Inner Mongolia. (German partner: The Leibniz Centre for Agricultural Landscape Research; Chinese partner: Institute of Agricultural Environment and Sustainable Development, Chinese Academy of Agricultural Sciences, Beijing).

(10) Influence of grazing pressure on the carbon isotope composition of the grassland of China: spatio-temporal variations at multiple scales. (German partner: Center of Life and Food Sciences Weihenstephan, Technical University of Munich; Chinese partner: Institute of Botany, Chinese Academy of Sciences, Beijing).

(11) Surface and satellite based remote sensing to infer rain rates within the Xilin catchment. (German partner: German Meteorological Service, Meteorological Observatory Lindenberg-Richard Aßmann Observatory; Chinese partner: Institute for Atmospheric Physics, Chinese Academy of Sciences and Chinese Meteorological Administration, Dept. of Observations and Telecommunication, Division of Upper Air Observations).

The present study was carried out within the frame of sub-project 4 of the Institute of Animal Nutrition and Physiology, Christian-Albrechts-Universitätzu Kiel. In close cooperation with the Institute of Crop Science, Christian-Albrechts-Universitätzu Kiel (sub-project 3), a field grazing trial had been established in 2005 on a *Leymus chinen-*

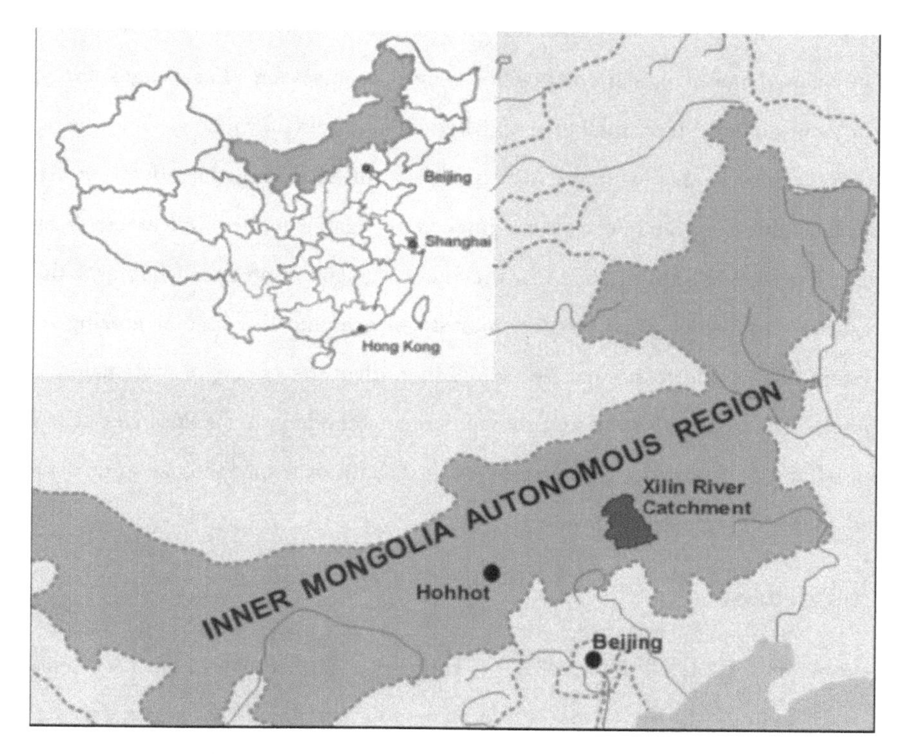

Figure 1.1 Location of the study site in the Xilin River Basin, Inner Mongolia, China

sis and *Stipagrandis* dominated plant community in the typical steppe of Inner Mongolia (Figure 1.1). The main aims of the Sub-projects 3 and 4 were to investigate the impacts of different grazing management systems and stocking rates on herbage species composition, herbage quality and quantity, digestibility of ingested herbage, feed intake, and live weight gain of sheep and to develop management concepts for a sustainable utilization of the grassland resources.

This dissertation focuses on the effects of grazing systems on digestibility of ingested organic matter, feed intake, and live weight gain of grazing sheep in the steppe of Inner Mongolia. Chapter 1 documents a general introduction for the utilization of the steppe in the Inner Mongolia, and hence, the negative effects of high stocking rates on herbage and animal production. Chapter 2 describes a study where the impacts of ROT and CON on herbage chemical composition, herbage mass on offer, digestibility of ingested organic matter, feed intakes, and live weight gain of grazing sheep were measured. Chapter 3 presents results of an experiment investigating the effects of ALT and CON on digestibility of organic matter, intakes of organic matter and metabolizable energy, and live weight gain of sheep in the Inner Mongolian steppe. Chapter 4 discusses the measured parameters which are appropriate to evaluate the special grazing management system effects and to summarize factors that affect the effects of systems on herbage and animals. The general discussion provides recommendations for approver grazing system applied in the Inner Mongolian steppe to maintain or even increase animal performance and long-term grassland productivity.

1.6 References

Ackermann, C. J., Purvis, H. T., Horn, G. W., Paisley, S. I., Reuter, R. R., Bodine, T. N., 2001. Performance of light *vs.* heavy steers grazing plains old world bluestem at three stocking rates. J. Anim. Sci.79, 493-499.

Aharoni, Y., Brosh, A., Orlov, A., Shargal, E., Gutman, M., 2004. Measurements of energy balance of grazing beef cows on Mediterranean pasture, the effects of stockingrate and season: 1. Digesta kinetics, faecal output and digestible dry matter intake. Livest. Sci.90, 89-100.

Allan, B. E., 1985. Grazing management on pasture and animal production from oversown tussock grassland. Proc. N. Z. Grassland Assoc.46, 119-125.

Allan, B. E., 1997. Grazing management of oversown tussock country 3. Effects on liveweight and wool growth of Merino wethers. N. Z. J. Agric. Res. 40, 437-447.

Biondini, M. E., Patton, B. D., Nyren, P. E., 1998. Grazing intensity and ecosystem processes in a northern mixed-grass prairie, USA. Ecol. Appl. 8, 469-479.

Briske, D. D., Derner, J. D., Brown, J. R., Fuhlendorf, S. D., Teague, W. R., Havstad, K. M., Gillen, R. L., Ash, A. J., Willms, W. D., 2008. Rotational grazing on rangelands: Reconciliation of perception and experimental evidence. Rangeland Ecol. Manage.61, 3-17.

Campbell, A. G., 1966. Grazed pasture parameters. II. Pasture dry-matter use in a stocking rate and grazing management experiment with dairy cows. J. Agri. Sci. 67, 211-216.

Chen, Z. Z., Wang, S. P., 2000. Typical steppe Ecosystem of China.2nd ed. Beijing: Science Press, p.9-45 (in Chinese).

Chen, Z. Z., Wang, S. P., 2006. Discussion on establishment of grassland ecological compensation mechanism. ActaAgrestia Sin.14, 1-3 (in Chinese).

Chen, Z. Z., Wang, S. P., Wang, Y. F., 2003. Update progress on grassland ecosystem research in Inner Mongolia steppe. Chinese Bulletin of Botany 20, 423-429 (in Chinese).

Christensen, L., Coughenour, M. B., Ellis, J. E., Chen, Z. Z., 2003. Sustainability of Inner Mongolian grasslands: Application of the Savanna model. J. Range Manage.56, 319-327.

Clarke, S. E., Tisdale, E. W., Skoglund, N. A., 1943. The effects of climate and grazing practices on short-grass prairie vegetation. C. Dept. Agr. Tech. Bull. 46, 53.

D'souza, G. E., Maxwell, E. W., Bryan, W. B., Prigge, E. C., 1990. Economic impacts of extended grazing systems. Am. J. Alternative Agric. 5, 120-125.

Derner, J. D., Hart, R. H., Smith, M. A., Waggoner, J. W., 2008. Long-term cattle gain responses to stocking rate and grazing systems in northern mixed-grass prairie. Livest. Sci.117, 60-69.

Ferraro, D. O., Oesterheld, M., 2002. Effect of defoliation on grass growth. A quantitative review. Oikos 98, 125-133.

Fynn, R. W. S., Connor, T. G. O', 2000. Effect of stocking rate and rainfall on rangeland dynamics and cattle performance in a semi – arid savanna, South Africa. J. Appl. Ecol.37, 491–507.

Garcia, F., Carrere, P., Soussana, J. F., Baumont, R., 2003. The ability of sheep at different stocking rates to maintain the quality and quantity of their diet during the grazing season. J. Agric. Sci.140, 113–124.

Gebremedhin, B., Pender, J., Tesfay, G., 2004. Collective action for grazing land management in crop–livestock mixed systems in the highlands of northern Ethiopia. Agric. Sys.82, 273–290.

Gerrish, J. R., 1993. Management intensive grazing: Principles and techniques. In: Proc Management Intensive Grazing Seminar, Dec 6, Rice, MN, p 30–49. University of Minnesota Ext Serv and Minnesota Institute for Sustainable Agric, St. Paul and Central Minnesota Forage Council, St. Cloud, MN.

Geus, J. De., 1950. A note on ration grazing in Holland. Grass Forage Sci.5, 251–252.

Glindemann, T., Wang, C. J., Tas, B. M., Schiborra, A., Gierus, M., Taube, F., Susenbeth, A., 2009. Impact of grazing intensity on herbage intake, composition, and digestibility and on live weight gain of sheep on the Inner Mongolian steppe. Livest. Sci.124, 142–147.

Grant, S. A., Barthram, G. T., Torvell, L., King, J., Elston, D. A., 1988. Comparison of herbage production under continuous stocking and intermittent grazing. Grass Forage Sci.43, 29–39.

Hafley, J. L., 1996. Comparison of Marshall and Surrey ryegrass for continuous and rotational grazing. J. Anim. Sci.74, 2269–2275.

Han, G., Li, B., Wei, Z., Li, H., 2000. Live weight change of sheep under 5 stocking rates in *Stipabreviflora* desert steppe. Grassl. China 38, 4–6 (in Chinese).

Heady, H. F., 1961. Continuous vs. specialized grazing systems: A review and application to the California annual type. J. Range Manage.14, 182–193.

Hodgson, J., 1979. Nomenclature and definitions in grazing studies. Grass Forage Sci.34, 11–18.

Hull, J. L., Meyer, J. H., Kromann, R., 1961. Influence of stocking rate on animal and forage production from irrigated pasture. J. Anim. Sci.20, 46–52.

Jamieson, W. S., Hodgson, J., 1979. The effect of daily herbage allowance and sward characteristics upon the ingestive behaviour and herbage intake of calves under strip–grazing management. Grass Forage Sci.34, 261–271.

Jiang, G. M., Han, X. G., Wu, J. G., 2006. Restoration and management of the Inner Mongolia grassland require a sustainable strategy. Ambio 35, 269–270.

Kang, L., Han, X. G., Zhang, Z. B., Sun, O. J., 2007. Grassland ecosystems in China: review of current knowledge and research advancement. Philosophical Transactions of the Royal Society B – Biological Sciences 362, 997–1008.

Kawamura, K., Akiyama, T., Yokota, H., Tsutsumi, M., Yasuda, T., Watanabe, O., Wang, S. P., 2005. Quantifying grazing intensities using geographic information systems and satellite remote sensing in the Xilingol steppe region, Inner Mongolia, China. Agric. Ecosyst. Environ.107, 83–93.

Kemp, D. R., Michalk, D. L., 2007. Towards sustainable grassland and livestock management. J. Agric. Sci.145, 543–564.

Kitessa, S. M., Nicol, A. M., 2001. The effect of continuous or rotational stocking on the intake and live–weight gain of cattle co–grazing with sheep on temperate pastures. Anim. Sci.72, 199–208.

Langlands, J. P., Bennett, I. L., 1973. Stocking intensity and pastoral production.2. Herbage intake of merino sheep grazed at different stocking rates. J. Agric. Sci.81, 205–209.

Li, B., 1999a. Grassland degradation and suggestions for strategic prevention in north China. ScientiaAgricultura Sin.30, 1-9 (in Chinese).

Li, B., 1999b. Steppe degradation in northern China and preventing measures. In: Xu, R. (ed.). Collected Papers of Li Bo. Science Press, Beijing. pp.153 (in Chinese).

Li, S. G., Harazono, Y., Oikawa, T., Zhao, H. L., He, Z. Y., Chang, X. L., 2000. Grassland desertification by grazing and the resulting micrometeorological changes in Inner Mongolia. Agric. For. Meteorol.102, 125-137.

Lin, L. J., Dickhoefer, U., Müller, K., Wang, C. J., Glindemann, T., Hao, J., Wan, H. W., Schönbach, P., Gierus, M., Taube, F., Susenbeth, A., 2012. Growth of sheep as affected by grazing system and grazing intensity in the steppe of Inner Mongolia, China. Livest. Sci.144, 140-147.

Liu, J., Zhang, Y., Li, Y., Wang, D., Han, G., Hou, F., 2008. Overview of grassland and its development in China. In: 2008 XXI International Grassland and VIII International Rangelands Congress Proceedings, Hohote, China. pp.3-10.

Long, G. A, 1986. Management of Grazing System, in: Joss, P. J., Lynch, P. W., and Williams, O. B. (Eds.), Rangelands: A resource under siege. Cambridge University Press, Cambridge, UK, pp.206-211.

Lu, C. D., 1988. Grazing behavior and diet selection of goats. Small Rumin. Res. 1, 205-216.

Lu, Z. J., Lu, X. S., Xin, X. P., 2005. Present situation and trend of grassland desertification of North China. ActaAgrestia Sin.13, 24-27 (in Chinese).

Manley, W. A., Hart, R. H., Samuel, M. J., Smith, M. A., Waggoner, J. W., Manley, J. T., 1997. Vegetation, cattle, and economic responses to grazing strategies and pressures. J. Range Manage.50, 638-646.

McMeekan, C. P., Walshe, M. J., 1963. The inter-relationship of grazing method and stocking rate in the efficiency of pasture utilization by dairy cattle. J. Agri. Sci.63, 147-166.

Merrill, L. B., 1954. A variation of deferred rotation grazing for use under southwest range conditions. J. Range Manage.7, 152–154.

Müller, K., Dickhoefer, U., Lin, L. J, Glindemann, T., Wang, C. J, Schönbach, P., Wan, H. W., Schiborra, A., Tas, B. M., Gierus, M., Taube, F., Susenbeth, A., 2012. Impact of grazing intensity on herbage quality, feed intake, and live weight gain of sheep grazing the steppe of Inner Mongolia, China. J. Anim. Sci. (submitted).

Owens, L. B., Shipitalo, M. J., 2009. Runoff quality evaluations of continuous and rotational over – wintering systems for beef cows. Agric. Ecosyst. Environ. 129, 482–490.

Popp, J. D., McCaughey, W. P., Cohen, R. D. H., 1997. Effect of grazing system, stocking rate and season of use on diet quality and herbage availability of alfalfa–grass pastures. Can. J. Anim. Sci.77, 111–118.

Ralphs, M. H., Kothmann, M. M., Taylor, C. A., 1990. Vegetation response to increased stocking rates in short – duration grazing. J. Range Manage. 43, 104–108.

Reardon, P. O., Merrill, L. B., 1976. Vegetative response under various grazing management systems in the Edwards Plateau of Texas. J. Range Manage. 29, 195–198.

Schiborra, A., 2007. Short–term effects of defoliation on herbage productivity and herbage quality in a semi – arid grassland ecosystem of Inner Mongolia, P. R. China. Ph. D. thesis, Christian–Albrechts–University of Kiel, Germany.

Schönbach, P., Wan, H. W., Schiborra, A., Gierus, M., Bai, Y. F., Müller, K., Glindemann, T., Wang, C. J., Susenbeth, A., Taube, F., 2009. Short–term management and stocking rate effects of grazing sheep on herbage quality and productivity of Inner Mongolia steppe. Crop Pasture Sci. 60, 963–974.

Schönbach, P., Wan, H. W., Gierus, M., Bai, Y. F., Müller, K., Lin,

L. J., Susenbeth, A., Taube, F., 2011. Grassland responses to grazing: effects of grazing intensity and management system in an Inner Mongolian steppe ecosystem. Plant Soil 340, 103–115.

Sun, H. L., 2005. Ecosystems of China. Science Press, Beijing, China.

Suttie, J. M., Reynolds, S. G., Batello, C., 2005. Grasslands of the world. Food and Agriculture Organization of the United Nations, Rome, Italy.

Thwaites, R., Terry, de L., Li, Y. H., Liu, X. H., 1998. Property rights, social change, and grassland degradation in Xilingol Biosphere Reserve, Inner Mongolia, China. Soc. Nat. Resour.11, 319–338.

Tong, C., Wu, J., Yong, S., Yang, J., Yong, W., 2004. A landscape-scale assessment of steppe degradation in the Xilin River Basin, Inner Mongolia, China. J. Arid Environ.59, 133–149.

VanPoollen, H. W., Lacey, J. R., 1979. Herbage response to grazing systems and stocking intensities. J. Range Manage.32, 250–253.

Virgona, J. M., Avery, A. L., Graham, J. F., Orchard, B. A., 2000. Effects of grazing management on phalaris herbage mass and persistence in summer–dry environments. Aust. J. Exp. Agric.40, 171–184.

Wang, C. J., Tas, B. M., Glindemann, T., Müller, K., Schiborra, A., Schoenbach, P., Gierus, M., Taube, F., Susenbeth, A., 2009. Rotational and continuous grazing of sheep in the Inner Mongolian steppe of China. J. Anim. Physiol. Anim. Nutr.93, 245–252.

Wang, D. L., Han, G. D., Bai, Y. G., 2005. Interactions between foraging behaviour of herbivores and grassland resources in the eastern Eurasian steppes. Plenary and invited papers from the XX International Grassland Congress, Dublin, Ireland. Grassland: A Global Resource. pp.97–110.

Wang, S. P., Li, Y. H., 1999. Degradation mechanism of typical grassland in Inner Mongolia. Chinese Journal of Applied Ecology 10, 437–441 (in Chinese).

Warner, J. R., Sharrow, S. H., 1984. Set stocking, rotational grazing and forward rotational grazing by sheep on western oregon hill pastures. Grass Forage Sci.39, 331-338.

Willms, W. D., Smoliak, S., Dormaar, J. F., 1985. Effects of stocking rate on a rough fescue grassland vegetation. J. Range Manage.38, 220-225.

Yiruhan, Hayashi, I., Nakamura, T., Shiyomi, M., 2001. Changes in floristic composition of grasslands according to grazing intensity in Inner Mongolia, China. Grassland Sci.47, 362-369 (Japanese).

Yu, M., Ellis, J. E., Epstein, H. E., 2004. Regional analysis of climate, primary production, and livestock density in Inner Mongolia. J. Environ. Qual.33, 1675-1681.

Zhang, Z. S., 2008. Main developments of national grassland works. Grassland Sci.25, 1-3 (in Chinese).

Zhao, H. L., Zhao, X. Y., Zhou, R. L., Zhang, T. H., Drake, S., 2005. Desertification processes due to heavy grazing in sandy rangeland, Inner Mongolia. J. Arid Environ.62, 309-319.

Zhao, W. Y., Li, J. L., Qi, J. G., 2007. Changes in vegetation diversity and structure in response to heavy grazing pressure in the northern Tianshan Mountains, China. J. Arid Environ.68, 465-479.

Chapter 2 Effects of rotational and continuous grazing on herbage quality, feed intake, and performance of sheep on a semi-arid grassland steppe

2 Effects of rotational and continuous grazing on herbage quality, feed intake, and performance of sheep on a semi-arid grassland steppe

2.1 Abstract

Compared to continuous (CON) grazing, rotational (ROT) grazing increases herbage production and thereby the resilience of grasslands to intensive grazing. Results on feed intake and animal performance, however, are contradictory. Hence, the objective was to determine the effects of ROT and CON on herbage mass (HM), digestibility of ingested organic matter (dOM), organic matter intake (OMI), and live weight gain (LWG) of sheep in the Inner Mongolian steppe, China. During June-September 2005-2008, 2 2-ha plots were used for each grazing system. At ROT, plots were divided into 4 0.5-ha paddocks that were grazed for 10 d each at a moderate stocking rate. Instead, CON sheep grazed the whole plots throughout the entire grazing season. In the beginning of every month, dOM was estimated from fecal crude protein concentration. Feces excretion was determined using titanium dioxide in 6 sheep per plot. The animals were weighed every month to determine their LWG. Across the years, HM did not differ between systems ($P=0.820$). However, dOM, OMI, and LWG were lower at ROT than at CON ($P\leq0.005$). Our study thus showed that ROT does not improve herbage growth, feed intake and performance of sheep and suggestsed that stocking rates rather than management system determine the ecological sustainability of pastoral livestock systems in semi-arid environments.

2.2 Introduction

Several grazing systems were developed to lower grazing pressure on the natural fodder resources without reducing animal performance. In a continuous grazing (CON) system, animals graze the same area during the whole vegetation period (Hodgson, 1979). Instead, the grazing area is divided into several paddocks that are grazed in sequence in the rotational grazing (ROT) system (Frame, 1992). Although ROT increases stocking densities in a short term, it includes resting periods when the vegetation is allowed to recover from grazing. This may maintain or even increase short-term as well as long-term grassland productivity (Virgona et al., 2000) and as a consequence, animal production (Allan, 1997). However, in a literature review Briske et al. (2008) found that 87% of the 32 considered studies on vegetation responses reported a similar or lower herbage production at ROT than at CON, whereas 92% of the 38 experiments that analyzed grazing system effects on animal performance determined a similar or lower animal live weight gain (LWG) at ROT than at CON. Hence, the effects of ROT on the quantity and quality of herbage and animal LWG are inconsistent. This may be, amongst other reasons, caused by differences in the studied ecosystem, the animal species used, and the applied stocking rates (Van Poollen and Lacey, 1979; Warner and Sharrow, 1984; Kitessa and Nicol, 2001). Moreover, Schönbach et al. (2009) who concluded that short-term experiments are inadequate to compare system effects when measuring the effects of different grazing systems on herbage quality and production in the Inner Mongolian steppe. Within the same research project, Wang et al. (2009a) suggested that very low precipitation might curtail the positive effects of ROT on herbage and animal performance. Their study was therefore continued to analyze the effects of ROT and CON on herbage as well as feed intake and LWG of sheep. We aimed to determine whether they depend on the duration of the experiment and differ between study years or throughout the grazing seasons due to changing climatic conditions.

2.3 Materials and methods

2.3.1 Site of study and current land use system

The study was conducted in the Inner Mongolian steppe (43°38′N, 116°42′E) at an approximate altitude of 1200 m above sea level. In the past 6 decades, human population as well as the number of grazing cattle and in particular sheep in this region rapidly increased, which strongly reduced the available grassland per animal (Jiang et al., 2006). Moreover, the sedentarization of the formerly nomadic pastoralists increased grazing pressure close to settlements, whereas distant grassland areas are nowadays only used for hay making. Although high grazing intensities could increase animal production per unit of land area in a short term (Glindemann et al., 2009b), this excessive utilization is not sustainable and might result in a long-term decline in ecosystem production (Christensen et al., 2003). The climate is a semi-arid, continental temperate steppe climate. Annual precipitation averages 343 mm and mean annual air temperature is 0.7℃ (climate data was collected at a weather station close to our experimental areas in 1982—2004). Monthly precipitations and mean monthly air temperatures in 2005—2008 are shown in Figure 2.1. Rainfall mainly occurs from June to August and the vegetation period lasts from April to September for about 150 to 180 d per year. The dominant soil type in the region is a chestnut soil (Chen and Wang, 2000). The steppe vegetation is dominated by *Leymus chinensis* Trin. and *Stipa grandis* P. A. Smirn (Bai et al., 2004). Vegetation cover is about 30%~40% and may reach 60%~70% in wet years (Chen and Wang, 2000). Annual yields of herbage mass (HM) range between 152 and 240 g dry matter (DM) /m^2 in 2005—2008 (Schönbach et al., 2011).

2.3.2 Experimental design

The experiments lasted from the mid June to the mid September (grazing season) for 98, 90, 93, and 95 days in 2005, 2006, 2007, and 2008, respectively. Two different grazing management systems were tested: a ROT and a CON system. The latter is similar with the current grazing system in Inner Mongolia (see previous section). Each system was tested on two permanently fenced plots: a flat and a moderately sloped plot to account for differences in the geographical setting. For ROT, plots (2 ha each) were divided into four equally sized paddocks that were grazed sequentially for 10 d each, followed by a resting period of 30 d (i. e. each paddock was grazed 2~3 times per grazing season). For CON, sheep were allowed to graze the whole plots throughout the grazing season. Outside of the grazing seasons, the plots remained ungrazed. 9 sheep per plot (i. e.4.5 sheep/ha) were used in 2005 and 2006, whereas sheep numbers were adjusted every month to HM on offer and the total live weight (LW) of all animals per plot in 2007 and 2008 to maintain a similar herbage allowance (HA) of 4.5~6 kg DM/kg LW across the grazing season. Across 2007 and 2008, the mean stocking rates were 4.0 sheep/ha at ROT and 3.9 sheep/ha at CON, respectively. This equals a moderate grazing intensity (4.5 sheep/ha) for the grasslands in this region (Wang et al., 2001). Moderate stocking rates were used in both systems to test whether in addition to reduced grazing intensities, ROT can contribute to a more sustainable grassland use in Inner Mongolia. Details on implemented stocking rates and resulting HA values during experimental years are given in Table 2.1.

2.3.3 Herbage mass and quality

After removing the litter, HM on offer was measured by cutting the sward to 1 cm above ground level in 2.0×0.25 m^2 frames (n = 3; representative areas were chosen to account for differences in the sward composition within each plot) at the beginning of June, July, August, and September each year. Mean HM on ROT and CON plots was (653±51), (442±136), (631±92), and (616±50) kg DM/ha in June 2005, 2006, 2007, and 2008, respectively. For ROT, samples were taken in grazed paddocks at the beginning of each 10-d grazing period and the mean HM on offer was calculated for every month and experimental year. The collected herbage material was oven-dried for 24 h at 60℃, weighed, and pooled by plot for milling. After grinding to pass a 1-mm mesh by a Cyclotec 1093 Sample Mill (Tecator, Sweden), herbage samples were analyzed for DM, crude protein (CP), neutral detergent fiber (NDF), acid detergent fiber (ADF), acid detergent lignin (ADL), cellulase digestible organic matter (CDOM), and metabolizable energy (ME) by near - infrared reflectance spectroscopy (NIRS). For details on calibration and validation of NIRS as well as the analytical procedures see Schönbach et al. (2009).

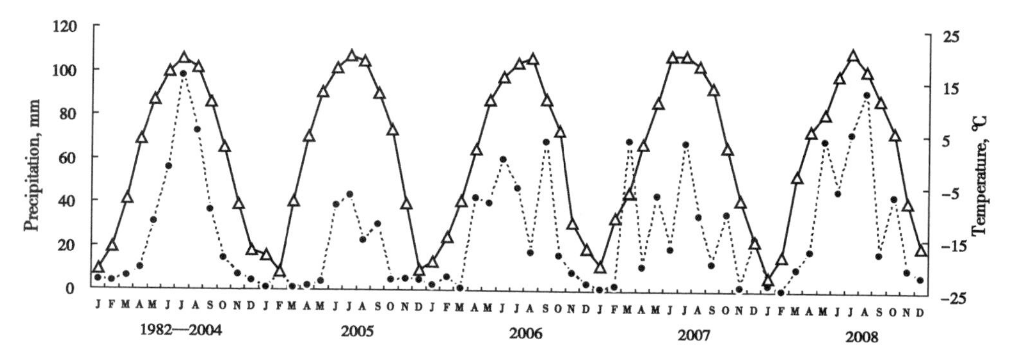

Figure 2.1 Mean monthly temperatures

(△, ——) and precipitations (●, ------) at the experimental site during 1982—2004 and in the study years 2005, 2006, 2007, and 2008.

2.3.4 Animals and live weight gain

In 2005, 2006, 2007, and 2008, 36, 36, 31, and 33 female sheep of the local fat-tailed breed were used. The animals were about 15 months old and neither pregnant nor lactating. Sheep were randomly assigned to the grazing plots after shearing and anthelmintic treatments. During the grazing seasons, they had free access to water and mineral lick stones and were allowed to graze for the whole day. All sheep were weighed on two consecutive days at the beginning of the grazing seasons, using an electronic platform scale (0.1-kg accuracy). Their initial LW was (31.1 ± 5.7), (31.4 ± 4.6), (31.1 ± 2.4), and (30.2 ± 4.6) kg in June 2005, 2006, 2007, and 2008, respectively. Thereafter, the animals were weighed again on the 11th and 12th of every month. The mean LW of the two days was used to calculate the animals' LWG during each month according to Equation 1:

Table 2.1 Mean live weight (LW) of sheep, stocking rate (SR), and herbage allowance (HA) on continuously (CON) and rotationally grazed (ROT) plots during the grazing seasons of 2005—2008 (n=2).

Parameter	GS	2005					2006					2007					2008				
		June	July	Aug.	Sept.	Mean	June	July	Aug.	Sept.	Mean	June	July	Aug.	Sept.	Mean	June	July	Aug.	Sept.	Mean
LW, kg/sheep	CON	30.0	33.1	34.9	36.3	33.6	31.6	33.6	37.1	37.5	35.0	30.7	34.1	36.5	41.3	35.7	30.3	36.9	39.9	44.1	37.8
	ROT	32.1	34.9	36.7	38.5	35.6	31.1	33.3	36.7	37.8	34.7	31.4	33.3	36.9	41.0	35.7	30.2	35.9	38.5	40.4	36.3
SR, sheep/ha	CON	4.5	4.5	4.5	4.5	4.5	4.5	4.5	4.5	4.5	4.5	3.5	4.0	4.3	3.8	3.9	3.8	3.8	3.8	3.8	3.8
	ROT	4.5	4.5	4.5	4.5	4.5	4.5	4.5	4.5	4.5	4.5	3.5	3.5	3.5	3.3	3.5	4.3	4.5	4.3	4.3	4.4
HA, kg DM/kg LW	CON	—	7.1	4.7	3.2	5.0	—	3.1	1.7	1.3	2.0	—	4.6	3.2	5.3	4.4	—	5.7	5.5	5.4	5.5
	ROT	—	4.9	2.5	3.5	3.6	—	4.0	3.7	1.7	3.1	—	4.5	5.4	3.1	4.3	—	3.4	4.2	6.2	4.6
HA, kg DM/ $kg^{0.75}$ LW	CON	—	16.6	11.3	7.6	11.8	—	7.2	4.2	3.1	4.8	—	11.0	7.8	13.4	10.7	—	14.1	13.8	13.8	13.9
	ROT	—	11.9	6.0	8.6	8.8	—	9.5	9.0	4.3	7.6	—	10.5	12.7	7.4	10.2	—	8.0	10.3	15.3	11.2

Aug., August; GS, grazing system; Sept., September.

$$LWG\ [g/d] = (LW_m\ [kg] - LW_{m-1}\ [kg]) / D_m \cdot 1\,000 \quad (1)$$

Where LWG is the daily LWG, LW the mean LW of the two weighings per month, and D the number of days between weighing dates. m and m−1 indicate two consecutive months (m=June, July, August, and September). Daily LWG per hectare (LWGh, g/ha) was calculated from the mean daily LWG of individual sheep (LWGs, g/d) and the stocking rate (sheep/ha) in the respective plot.

2.3.5 Feed intake and digestibility of ingested herbage

At the beginning of the grazing seasons, six sheep per plot were randomly chosen to determine organic matter intake (OMI) and digestibility of ingested herbage organic matter (dOM). Daily OMI of sheep was calculated from dOM and daily fecal organic matter excretion (g/d). The latter was measured using the external marker titanium dioxide (TiO_2). For this, one gelatine capsule filled with 2.5 g TiO_2 was orally administered to the 6 sheep per plot about 10: 30am each day for the first ten days of July, August, and September, respectively. Immediately after marker application, fecal samples (approximately 25 g fresh matter per sheep) were taken from the rectum on days 6 to 10 (sampling period) and frozen. At the end of each sampling period, all fecal samples were thawed and pooled to one sample per sheep (150 g fresh matter). The pooled samples were divided into two sub-samples. One sub-sample was oven-dried at 60℃ for 48 h and used to determine fecal TiO_2 concentration analysis according to Glindemann et al. (2009a) using the marker recovery rate of 100%. The other sample was analyzed for DM, crude ash (CA), and nitrogen (N) concentrations according to the methods of the Chinese Technical Committee for Feed Industry Standardization and the Chinese Association of Feed Industry (2000). Briefly, DM is determined by drying the samples at 105 ℃ for 24 h, CA by incineration at 550 ℃, and N by the Kjeldahl procedure.

According to the equation suggested by Wang et al. (2009b), dOM was calculated as follows (Equation 2):

$$dOM\ [\%] = 89.9 - 64.4 \cdot \exp\ (-0.5774 \cdot \text{fecal CP}\ [\text{g/kg organic matter}]\ /100) \quad (2)$$

Where dOM is the digestibility of ingested organic matter and CP the CP concentration in fecal organic matter (= N × 6. 25). Subsequently, OMI was estimated by equation 3:

$$OMI\ [\text{g/d}] = \text{fecal organic matter}\ [\text{g/d}]\ /\ (100 - dOM\ [\%]) \quad (3)$$

Where OMI is the daily OMI of sheep and dOM the digestibility of ingested organic matter.

Intakes of digestible organic matter (DOMI) and ME (MEI) were calculated by multiplying OMI by dOM or ME concentrations, respectively. The latter was estimated according to equation 4 derived by Aiple (personal communication) on the basis of the data published by Aiple et al. (1992):

$$ME\ [\text{MJ/kg organic matter}] = -0.9 + 0.170 \cdot dOM\ [\%] \quad (4)$$

Where ME is the dietary ME concentration and dOM the digestibility of ingested organic matter.

2.3.6 Statistical analysis

Mean values of the 6 selected sheep per plot were used for statistical analyses. Hence, in total, 48 observations (2 systems · 4 years · 3 months · 2 plots) were obtained for all animal and herbage parameters. Data were analyzed using the Mixed Model procedure of SAS version 9.1 (SAS Institute Inc., Cary, NC, USA) to test the effect of grazing system, year, and month on HM, herbage composition, dOM, OMI, DOMI, MEI, and LWG, respectively. The following model was used:

$$y_{ijkl} = \mu + GS_i + YE_j + M_k + (GS \times YE)_{ij} + (GS \times M)_{ik} + (YE \times M)_{jk} + (GS \times YE \times M)_{ijk} + P_l + e_{ijkl} \quad (5)$$

Where y is the dependent variable, μ the overall mean, GS_i grazing management system (ROT and CON), YE_j the year (2005, 2006, 2007, and 2008), M_k the month (July, August, and September), P_l the plot (flat and sloped), and e_{ijkl} the random experimental error. Year was treated as repeated measurement. All other factors and their interactions were treated as fixed effects. Probabilities for all effects and their interactions were determined. When effects were significant ($P \leqslant 0.05$), the Tukey test was used for pair wise comparisons of least squares means.

2.4 Results

2.4.1 Herbage mass and quality

The effects of grazing system on HM and herbage quality are shown in Table 2.2. Across the four study years, mean HM on offer at ROT and CON was 641 and 628 kg DM/ha, respectively, and did not differ between grazing systems ($P=0.820$). Chemical composition of herbage was influenced by grazing system. Concentrations of CP ($P<0.001$), CDOM ($P=0.034$), and ME ($P=0.019$) were lower at ROT than at CON, whereas concentrations of ADF were greater at ROT (33.4 % DM) than at CON (32.8% DM, $P=0.025$). Grazing system did not affect concentrations of NDF and ADL ($P \geqslant 0.257$). Moreover, there were no significant interactions between grazing system and year ($P \geqslant 0.068$) and grazing system and month ($P \geqslant 0.304$) for herbage mass and quality parameters.

Mass and chemical composition of herbage on offer strongly differed between years ($P \leqslant 0.001$). Whereas HM on offer was lowest in 2006 (405 kg DM/ha, $P \leqslant 0.036$), it reached 665, 640, and 828 kg DM/ha in 2005, 2007, and 2008, respectively. In contrast thereto, concentrations of ME ($P \leqslant 0.001$) and CDOM ($P \leqslant 0.051$) were highest in 2006, whereas NDF ($P \leqslant 0.085$) and ADF ($P \leqslant 0.149$) contents were lowest in this year.

Table2.2 Mean herbage mass (HM) and chemical composition on continuously (CON) and rotationally grazed (ROT) plots during the grazing seasons of 2005—2008 (least squares means; n=2).

Parameter	Year	CON				ROT				Mean	SEM	*P*-value						
		July	Aug.	Sept.	Mean	July	Aug.	Sept.	Mean			GS	YE	M	GS×YE	GS×M	YE×M	GS× YE×M
HM, kg DM/ha	2005	999	716	499	738	769	406	603	592	665^{B}	34.3	0.820	0.001	0.836	0.268	0.550	0.013	0.166
	2006	450	283	210	314	590	601	293	494	404^{A}								
	2007	529	494	870	631	561	797	585	648	640^{AB}								
	2008	786	823	901	837	544	779	1 161	828	832^{B}								
CP, % DM	2005	10.1	8.7	8.2	9.0	9.3	8.5	6.6	8.1	8.6^{A}	0.34	< 0.001	< 0.001	< 0.001	0.419	0.660	0.129	0.385
	2006	15.3	14.4	12.5	14.0^{b}	13.4	10.8	11.0	11.7^{a}	13.0^{C}								
	2007	12.7	12.7	10.4	11.9	11.5	12.6	8.9	11.0	11.5^{B}								
	2008	13.9	13.9	12.9	13.6	13.5	10.9	12.9	12.4	13.0^{C}								
NDF, % DM	2005	72.5	72.4	72.2	72.4	72.4	74.1	73.3	73.3	72.8^{C}	0.36	0.442	< 0.001	0.044	0.557	0.369	0.094	0.086
	2006	65.6	69.0	68.8	67.8	66.9	69.4	65.6	67.3	67.5^{A}								
	2007	69.2	71.6	70.1	70.3	70.6	70.8	70.6	70.6	70.5^{B}								
	2008	68.6	69.1	67.0	68.2	70.0	67.4	68.4	68.6	68.4^{A}								
ADF, % DM	2005	33.6	34.2	34.1	33.9^{a}	34.4	35.3	37.0	35.5^{b}	34.7^{C}	0.22	0.025	< 0.001	0.013	0.068	0.967	0.032	0.034
	2006	31.0	32.0	33.1	32.0	31.5	33.4	31.1	32.0	32.0^{A}								
	2007	32.2	32.8	34.7	33.2	32.7	32.8	34.0	33.2	33.2^{B}								
	2008	32.6	32.1	31.6	32.1	33.5	31.9	33.4	32.9	32.5^{AB}								

(attached table)

Parameter	Year	CON				ROT				Mean	SEM	P-value						
		July	Aug.	Sept.	Mean	July	Aug.	Sept.	Mean			GS	YE	M	GS×YE	GS×M	YE×M	GS×YE×M
ADL, % DM	2005	4.1	5.0	5.0	4.7^{a}	4.3	5.4	5.7	5.1^{b}	4.9^{AB}	0.08	0.257	0.001	< 0.001	0.081	0.304	0.007	0.251
	2006	3.8	5.3	5.1	4.7	3.8	4.8	5.4	4.7	4.7^{A}								
	2007	4.7	5.1	5.8	5.2	4.9	5.2	5.4	5.1	5.2^{B}								
	2008	4.3	4.9	4.7	4.6	4.4	4.7	5.0	4.7	4.7^{A}								
CDOM, % organic matter	2005	60.1	57.8	56.3	58.1	58.4	54.6	54.1	55.7	56.9^{A}	0.46	0.034	< 0.001	< 0.001	0.109	0.458	0.078	0.726
	2006	66.9	62.0	60.8	63.2^{b}	63.4	59.4	61.6	61.5^{a}	62.3^{C}								
	2007	61.7	58.9	56.0	58.9	62.5	59.9	57.7	60.0	59.5^{B}								
	2008	62.7	60.2	62.1	61.7	59.4	60.0	60.6	60.0	60.8^{BC}								
ME, MJ/kg DM	2005	8.3	7.9	7.7	8.0	8.1	7.4	7.3	7.6	7.8^{A}	0.07	0.019	< 0.001	< 0.001	0.120	0.474	0.061	0.614
	2006	9.4	8.7	8.5	8.9	9.0	8.4	8.7	8.7	8.8^{C}								
	2007	8.5	8.3	7.9	8.2	8.6	8.4	8.0	8.3	8.3^{B}								
	2008	8.7	8.3	8.4	8.5^{b}	8.1	8.2	8.3	8.2^{a}	8.3^{B}								

[a,b]Within rows means without a common superscript differ ($P \leqslant 0.05$) between means for each GS and year. [A,B,C] Within columns means without a common superscript differ ($P \leqslant 0.05$) between the overall means for each years. ADF, acid detergent fiber; ADL, acid detergent lignin; Aug., August; CDOM, cellulase digestible organic matter; CP, crude protein; DM, dry matter; GS, grazing system; M, month; ME, metabolizable energy; NDF, neutral detergent fiber; SEM, standard error of the mean; Sept., September; YE, year.

Herbage quality differed between months. Concentrations of CP, CDOM, and ME were greater in July with 12.4 % DM, 61.9 % organic matter, and 8.6 MJ/kg DM in than in September with 10.4 % DM, 58.7 % organic matter, and 8.1 MJ/kg DM in September ($P< 0.001$ for all parameters). In contrast thereto, concentrations of ADF and ADL were greater in July (32.7 % DM, $P=0.013$; 4.3 % DM, $P< 0.001$) than in September (33.6 % DM; 5.3 % DM). Across the four study years, HM did not differ between months ($P=0.836$).

2.4.2 Digestibility of ingested herbage and feed intake

Across the four study years, dOM, OMI, DOMI, and MEI were lower at ROT than at CON ($P \leqslant 0.002$) and there were no significant interactions between grazing system and year for these parameters ($P \geqslant 0.384$; Table 2.3). Nevertheless, the differences in dOM between the two systems were only significant in 2006 ($P=0.029$) and 2007 ($P=0.001$). On the contrary, OMI, DOMI, and MEI were lower at ROT than at CON in 2005 ($P \leqslant 0.045$ for all parameters) and 2006 ($P=0.007$, $P=0.002$, $P=0.002$), whereas no differences were found in 2007 ($P \geqslant 0.070$) and 2008 ($P \geqslant 0.189$).

Moreover, dOM ($P< 0.001$), OMI ($P=0.018$), DOMI ($P=0.001$), and MEI ($P=0.002$) differed between years. With the preceding study years, dOM increased from 0.55 in 2005 to 0.61 in 2008. Lowest OMI, DOMI, and MEI were observed in 2005 and 2007, whereas respective values were the highest in 2006 and 2008.

Across the study years, dOM ($P< 0.001$), OMI ($P=0.028$), DOMI ($P=0.003$), and MEI ($P=0.003$) were greater in July than in September, which was due to significantly greater digestibility and intake values determined in July 2006 ($P \leqslant 0.047$ for all parameters). No differences between months were determined in any of the other years, except for dOM, which was greater in July and August than in September in 2007 ($P \leqslant 0.034$).

2.4.3 Live weight gain

Across the study years, LWGs was lower at ROT (80 g/d) than at CON (104 g/d; $P=0.005$). However, there was a tendency of an interaction between grazing system and year ($P=0.051$), the difference was only due to a lower LWGs of ROT sheep in 2008 ($P=0.013$; Table 2.3). On the contrary, grazing system did not affect LWGh across the study years ($P=0.248$). A significant effect of the interaction between grazing system and month on LWGs was found ($P=0.036$) as LWGs was greater at CON than at ROT in July ($P=0.023$) but not in August ($P=0.999$) and September ($P=0.253$).

Table2.3 Mean digestibility of ingested organic matter (dOM), daily organic matter intake (OMI), digestible OMI (DOMI), and metabolizable energy intake (MEI) as well as daily live weight gain of sheep (LWGs) and per hectare (LWGh) at rotationally (ROT) and continuously grazed (CON) plots during the grazing seasons of 2005—2008 (least squares means; n=2).

Parameter	Year	CON				ROT				Mean	SEM	*P*-value						
		July	Aug.	Sept.	Mean	July	Aug.	Sept.	Mean			GS	YE	M	GS×YE	GS×M	YE×M	GS×YE×M
dOM	2005	0.57	0.55	0.54	0.55	0.54	0.56	0.52	0.54	0.55^{A}	0.01	< 0.001	< 0.001	< 0.001	0.384	0.425	< 0.001	0.016
	2006	0.62	0.58	0.55	0.59^{b}	0.58	0.58	0.54	0.57^{a}	0.58^{B}								
	2007	0.57	0.63	0.58	0.59^{b}	0.56	0.57	0.55	0.56^{a}	0.58^{B}								
	2008	0.63	0.61	0.60	0.61	0.60	0.59	0.61	0.60	0.61^{C}								
OMI, $g/kg^{0.75}$ LW	2005	79.8	84.2	76.0	80.0^{b}	71.4	63.9	72.2	69.2^{a}	74.6^{A}	3.68	0.002	0.018	0.007	0.512	0.665	0.402	0.704
	2006	106	85.9	90.1	94.0^{b}	89.4	66.4	79.3	78.4^{a}	86.2^{B}								
	2007	82.0	75.8	77.4	78.4	79.4	74.2	62.1	71.9	75.1^{A}								
	2008	92.1	79.3	75.9	82.4	83.5	69.7	78.4	77.2	79.8^{AB}								
DOMI, $g/kg^{0.75}$ LW	2005	45.0	46.6	40.5	44.1^{b}	38.4	34.7	37.9	36.9^{a}	40.5^{A}	2.15	< 0.001	0.001	0.001	0.442	0.506	0.130	0.634
	2006	66.2	50.1	50.0	55.3^{b}	52.1	38.1	42.4	44.2^{a}	49.8^{C}								
	2007	46.5	47.6	44.8	46.3	44.7	42.2	34.3	40.4	43.3^{AB}								
	2008	58.1	48.1	46.9	50.6	50.1	41.1	48.1	46.5	48.5^{BC}								
MEI, $MJ/kg^{0.75}$LW	2005	0.69	0.72	0.63	0.68^{b}	0.59	0.55	0.57	0.57^{a}	0.62^{A}	0.03	< 0.001	0.002	0.002	0.473	0.481	0.135	0.640
	2006	1.03	0.77	0.77	0.86^{b}	0.81	0.59	0.65	0.68^{a}	0.77^{C}								
	2007	0.72	0.74	0.69	0.72	0.69	0.65	0.53	0.62	0.66^{AB}								
	2008	0.91	0.75	0.71	0.79	0.78	0.64	0.75	0.72	0.75^{BC}								

(attached table)

Parameter	Year	CON				ROT				Mean	SEM	*P*-value						
		July	Aug.	Sept.	Mean	July	Aug.	Sept.	Mean			GS	YE	M	GS×YE	GS×M	YE×M	GS×YE×M
LWGs, g	2005	94	64	39	65	85	65	48	66	66^{A}	10	0.005	< 0.001	0.007	0.051	0.036	< 0.001	0.363
	2006	114	114	17	81	111	112	-6	72	77^{A}								
	2007	145	90	154	130	71	113	130	104	117^{B}								
	2008	173	97	151	140^{b}	75	95	68	79^{a}	109^{B}								
LWGh, g/ha	2005	423	286	174	294	382	293	215	296	295^{A}	30	0.248	0.001	0.016	0.412	0.032	<0.001	0.349
	2006	513	511	75	366	497	503	-27	324	345^{A}								
	2007	507	380	615	500	277	630	637	514	507^{B}								
	2008	654	355	565	525	352	466	339	385	455^{AB}								

^{a,b} Within rows means without a common superscript differ ($P \leqslant 0.05$) between means for each GS and year. ^{A,B,C} Within columns means without a common superscript differ ($P \leqslant 0.05$) between the overall means for each year. Aug., August; GS, grazing system; LW, live weight; M, month; SEM, standard error of the mean; Sept., September; YE, year.

Both, LWGs and LWGh differed between years ($P < 0.001$, $P = 0.001$) and months ($P = 0.007$, $P = 0.016$) and significant interactions between year and month were found ($P < 0.001$ for all parameters). Hence, LWGs and LWGh were lower in 2005 ($P \leqslant 0.003$, $P \leqslant 0.004$) and 2006 ($P \leqslant 0.031$, $P \leqslant 0.036$) than in 2007 and 2008. Across the study years, LWGs and LWGh were greater in July than in September ($P = 0.005$, $P = 0.018$), which was due to significant differences between the two months in 2006 ($P = 0.001$, $P = 0.001$). Instead, no differences between July and September in LWGs and LWGh were found in any of the other study years ($P \geqslant 0.266$).

2.5 Discussion

2.5.1 Effect of grazing system

In contrast to CON, ROT allows the vegetation to recover from animal grazing, thereby increases vegetation cover, avoids water and soil erosion, and reduces the risk of grassland degradation (Teague and Dowhower, 2003). It may thus offer a valuable management strategy for the sustainable use of the Inner Mongolian steppe for pastoral livestock production (Su et al., 2005). Hence, Virgona et al. (2000) reported that ROT increased the phalaris herbage mass compared to CON under sheep grazing. Moreover, Han et al. (1990) determined a higher nutritional quality of herbage at ROT than at CON in the Inner Mongolian steppe, although the authors did not find a positive effect on HM. In contrast thereto, Walker et al. (1989) studied the effect of management system on herbage in Southe United States and reported that ROT by cattle did not increase herbage quality. Similarly, Popp et al. (1997) found no difference in CP, CDOM, and ME concentrations between the two grazing systems for an alfalfa-grass pasture grazed by cattle in Southern United States. In our study, we also did not observe a positive effect of ROT on HM on offer. Moreover, CON not ROT increased concentrations of CP, CDOM, and ME across the four study years. This confirms the findings of Sharrow (1983) who studied the effect of sheep grazing on an annual grass-clover sward in Western United States during the dry summer period. Kitessa and Nicol (2001) who studied LWG of cattle and sheep co-grazing temperate pastures found that LWG of cattle was lower at CON than at ROT, whereas LWG of sheep was similar at both grazing systems. The authors suggested that this was due to the greater ability of sheep to select for a higher quality diet. Moreover, Warner and Sharrow (1984) stated that ROT is only superior to CON at a relatively high stocking rate (in this study a minimum of five ewes and nine lambs were used in 0.8 ha plots). Hence, the reasons for the contradicting study results might be differences in the climatic and environmental conditions, the animal species used as well as the applied stocking rates (Warner and Sharrow, 1984; Kitessa and Nicol, 2001; Bailey and Brown, 2011).

Stocking densities at ROT paddocks during the 10-d grazing periods were greater than at CON plots. This might have limited the ability of ROT sheep to select for preferred plant species or plant parts (Stuth et al., 1987), thereby reducing the nutritional quality of the animals' diets (Vallentine, 2001). Hence, together with the lower herbage quality, this might explain the lower dOM and consequently, the lower OMI, DOMI, and MEI of sheep at ROT than at CON. Results confirm those obtained by Hafley (1996) in a study with cattle grazing a ryegrass pasture in Southern United States. In accordance to the lower digestibility and intake values, LWGs was lower in ROT than in CON sheep. This did not lead to a lower LWGh due to higher stocking rates at ROT than at CON in 2008 (4.4 vs. 3.8 sheep/ha). However, it complies with results of Derner et al. (2008) who reported that ROT reduces LWG of cattle by 6% compared to CON. Other studies indicated that grazing system has no effect on animal LWG (Gammon, 1987; Manley et al., 1997; Wang et al., 2009a). In a literature review, Briske et al. (2008) showed that 92% of the grazing studies (n = 38) reported a similar or greater LWGs at CON than at ROT, whereas 84 % of them (n = 32) determined a similar or greater LWG per unit of land area. Hence, it appears that animal performance is primarily determined by grazing intensity rather than by grazing system (Heitschmidt et al., 1987; Manley et al., 1997; McCollum et al., 1999;

Derner et al., 2008) and that it might even be lower in ROT systems due to a poorer quality of herbage.

2.5.2 Effect of year

The difference in LWGs at CON or ROT was pronounced in 2008, whereas no significant differences between grazing systems were found during the other study years. In 2008, precipitation during the grazing period was 297 mm and higher than in 2005 (138 mm) and 2007 (178 mm). However, the lack of continuous forage removal as well as the maturation of plants during the resting periods of 30 d at ROT plots might have limited the positive effect of rainfall on re-growth and thus nutritional quality of herbage. Instead, herbage offered to CON sheep might have been less mature. Hence, herbage ME concentrations were lower at ROT than at CON, which might partly explain the lower LWGs of ROT sheep. Although the differences in dOM, DOMI, and MEI were not significant, mean MEI was 1.5 MJ/d lower at ROT than at CON in 2008 (Figure2.2). According to Jeroch et al. (1996), sheep need on average 23 kJ ME for each gram of LWG at 35 kg LW. The additional MEI of CON sheep would therefore have allowed for an extra gain of 65 g/d, which is similar to the measured difference in animal performance (61 g/d).

Compared with 2008, a similar precipitation was recorded in the grazing season of 2006 (233 mm). In this study year, CP and CDOM concentrations as well as dOM, OMI, DOMI, and MEI were significantly lower at ROT than at CON. Nevertheless, LWGs was similar at both systems. HM as well as HA were lowest in 2006, probably due to a carry-over effect of the very low rainfall in 2005. The lower HA might have increased the animals' physical activity and hence, their daily energy expenditures (Lin et al., 2011). The proportion of the MEI used for purposes other than maintenance and LWG (ME_x) can be estimated by deducting the animals' ME requirements for maintenance (ME_m) and gain (ME_g) from their daily MEI. Growing sheep require 430 kJ $ME/kg^{0.75}$ LW for maintenance and at 35 kg LW, 23 kJ ME for each gram of LWG

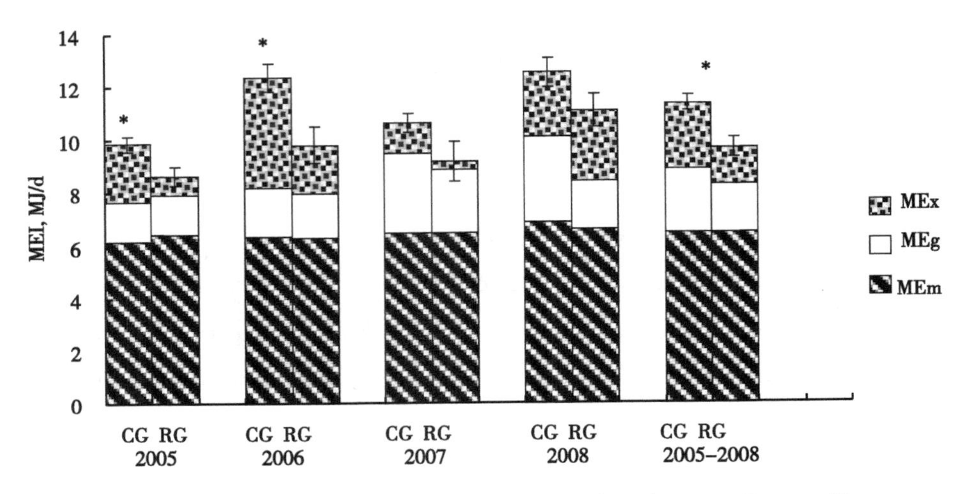

Figure2.2 Daily intakes of metabolizable energy (ME) and ME expenditures formaintenance (MEm), live weight gain (MEg), and purposes other thanmaintenance and gain (MEx) of sheep at continuous (CON) and rotational (ROT) grazing in 2005—2008

MEI [in MJ/d] was calculated by multiplying the mean MEintake [in MJ/kg live weight] of sheepin each year (see table 2.3) by the animals' respective metabolic live weight [$kg^{0.75}$ live weight] (see Table 2.1). * indicate significant ($P \leqslant 0.05$) differences between the MEI of CON and ROT sheep. Bars indicate standard deviations from the mean ME intake.

(Jeroch et al., 1996). Hence, ME_m, ME_g, and ME_x requirements of ROT sheep in our study were 6.45 MJ, 1.85 MJ, and 1.37 MJ/d, respectively, whereas at CON sheep required 6.46 MJ, 2.40 MJ, and 2.49 MJ/d for ME_m, ME_g, and ME_x. The differences in ME_x between CON and ROT were 1.5, 2.4, 0.9, and−0.2 MJ ME/d in the respective study years and were significant in 2005 ($P = 0.034$) and 2006 ($P = 0.002$), but not in the following two years ($P = 0.909$; $P = 0.997$). This seems to explain why LWGs was similar at ROT than at CON in the dry year 2005 and in 2006 when HA was very low, despite lower die digestibility and intake values.

2.5.3 Effect of month

Herbage quality decreased with proceeding vegetation period, which is consistent with findings of previous studies conducted within the frame of the same research project (Schönbach et al., 2009). As a consequence, dOM, feed intake, and LWGs decreased from July to September. However, in 2006, dOM, MEI, and LWGs across both systems were lower in September than in July, whereas the differences were not significant in 2005, 2007 (except LWGs), and 2008. This was most likely due to greater HA in these years (see previous section) which enabled the animals to compensate for the negative effects of herbage maturation by selecting for plants or plant parts of higher nutritional quality towards the end grazing season.

If ROT is only beneficial at high stocking rate and low HA (Warner and Sharrow 1984), one would expect that feed intake and performance of animals might be greater at ROT than at CON towards the end of the grazing season when HM and herbage quality decline. An interaction between grazing system and month was observed for LWGs. Across the four study years, LWGs of CON sheep was lower in September than in July, whereas no difference in animal performance was found between the two months at ROT. The latter was mainly due to the LWGs of ROT sheep in 2007 that was nearly twice as high in September than in July (130 g/d vs. 71 g/d), indicating a compensatory growth of sheep in this study year. In none of the other study years, ROT was superior to CON towards the end of the grazing season. Hence, the pronounced decrease in herbage quality found in our study did not alter the system effect. This suggests that the moderate grazing intensity used in our study, the adjustment of stocking rates to HM on offer in 2007 and 2008, and the similar decreases in herbage quality at both systems might have hampered the beneficial effect of ROT on feed intake and LWGs at the end of the grazing season.

2.6 Conclusions

Results of four study years showed that herbage quality, feed intake, and LWG of sheep grazing in the Inner Mongolian steppe at a moderate grazing intensity is inferior at ROT than at CON. However, higher energy intake in CON did not result in a corresponding increase in LWG, because animals used a greater amount of ingested energy for purposes other than maintenance and growth, likely for grazing and walking. The differences between grazing systems may vary between years according to the amount of rainfall and thus herbage mass on offer. Pronounced decreases in herbage quality with advancing vegetation period due to plant maturation do not alter the system effects. The lack of herbage re-growth during grazing seasons, the animals' selective feeding behavior, and the moderate grazing intensity appear to hamper any beneficial effect of ROT on pasture vegetation and sheep performance. Hence, further studies are required to test whether a ROT system is superior at grazing intensities to those used in our study.

2.7 References

Aiple, K. P., Steingass, H., Menke, K. H., 1992. Suitability of a buffered fecal suspension as the inoculum in the Hohenheim Gas Test.1. Modification of the method and its ability in the prediction of organic-matter digestibility and metabolizable energy content of ruminant feeds compared with rumen fluid as inoculum. J Anim. Physiol. Anim. Nutr.67, 57-66.

Allan, B. E., 1997. Grazing management of oversown tussock country 3. Effects on liveweight and wool growth of Merino wethers. N. Z. J. Agric. Res. 40, 437-447.

Bai, Y., Han, X., Wu, J., Chen, Z., Li, L., 2004. Ecosystem stability and compensatory effects in the Inner Mongolia grassland. Nature. 431, 181-184.

Bailey, D. W., Brown, J. R., 2011. Rotational grazing systems and livestock grazing behavior in shrub-dominated semi-arid and arid rangelands. Rangeland Ecol. Manage.64, 1-9.

Briske, D. D., Derner, J. D., Brown, J. R., Fuhlendorf, S. D., Teague, W. R., Havstad, K. M., Gillen, R. L., Ash, A. J., Willms, W. D., 2008. Rotational grazing on rangelands: Reconciliation of perception and experimental evidence. Rangeland Ecol. Manage.61, 3-17.

Chen, Z. Z., Wang, S. P., 2000. Typical steppe Ecosystem of China. 2nd ed. Beijing: Science Press, p.9-45 (in Chinese).

Chinese Technical Committee for Feed Industry Standardization and the Chinese Association of Feed Industry.2000. Chinese Technical Committee for Feed Industry Standardization and Chinese Association of Feed Industry. 1st ed. Beijing: China Standard Press (in Chinese).

Christensen, L., Coughenour, M. B., Ellis, J. E., Chen, Z. Z., 2003. Sustainability of Inner Mongolian grasslands: Application of the Savanna model. J. Range Manage.56, 319-327.

Derner, J. D., Hart, R. H., Smith, M. A., Waggoner, J. W., 2008. Long-term cattle gain responses to stocking rate and grazing systems in northern mixed-grass prairie. Livest Sci.117, 60-69.

Frame, J., 1992. Improved grassland management. 1st ed. Ipswich, U. K. Alexandria Bay, NY: Farming Press.

Gammon, D. M., 1987. A review of experiments comparing systems of grazing management on natural pastures. Proceedings of the Grassland Society of South Africa 13, 75-82.

Glindemann, T., Tas, B. M., Wang, C. J., Alvers, S., Susenbeth, A., 2009a. Evaluation of titanium dioxide as an inert marker for estimating faecal excretion in grazing sheep. Anim. Feed Sci. Technol.152, 186-197.

Glindemann, T., Wang, C. J., Tas, B. M., Schiborra, A., Gierus, M., Taube, F., Susenbeth, A., 2009b. Impact of grazing intensity on herbage intake, composition, and digestibility and on live weight gain of sheep on the Inner Mongolian steppe. Livest. Sci.124, 142-147.

Hafley, J. L., 1996. Comparison of Marshall and Surrey ryegrass for continuous and rotational grazing. J. Anim. Sci.74, 2269–2275.

Han, G. D., Xu, Z. X., Zhang, Z. T., 1990. Comparison of rotation and continuous–season grazing systems. J. Arid Land Resour. Envion.3, 355–362.

Heitschmidt, R. K., Dowhower, S. L., Walker, J. W., 1987. Some effects of a rotational grazing treatment on quantity and quality of available forage and amount of ground litter. J. Range Manage.40, 318–321.

Hodgson, J., 1979. Nomenclature and definitions in grazing studies. Grass Forage Sci.34, 11–18.

Jeroch, H., Kirchgeßner, M., Pallauf, J., Pfeffer, E., Schulz, E., Staudacher, W., 1996. Mitteilungen des Ausschusses für Bedarfsnormen der Gesellschaft für Ernährungsphysiologie – Energie – Bedarf von Schafen. In: Giesecke, D. (Ed.), Proceedings of the Society of Nutrition Physiology. Frankfurt: DLG–Verlag, p. 149–151 (in German).

Jiang, G. M., Han, X. G., Wu, J. G., 2006. Restoration and management of the Inner Mongolia grassland require a sustainable strategy. Ambio 35, 269–270.

Kitessa, S. M., Nicol, A. M., 2001. The effect of continuous or rotational stocking on the intake and live–weight gain of cattle co–grazing with sheep on temperate pastures. Anim. Sci.72, 199–208.

Lin, L. J., Dickhoefer, U., Müller K., Wurina, Susenbeth A., 2011. Grazing behavior of sheep at different stocking rates in the Inner Mongolian steppe, China. Appl. Anim. Behav. Sci.129, 36–42.

Manley, W. A., Hart, R. H., Samuel, M. J., Smith, M. A., Waggoner, J. W., Manley, J. T., 1997. Vegetation, cattle, and economic responses to grazing strategies and pressures. J. Range Manage.50, 638–646.

McCollum, F. T., Gillen, R. L., Karges, B. R., Hodges, M. E., 1999. Stocker cattle response to grazing management in tallgrass prairie. J. Range Manage.52, 120–126.

Popp, J. D., McCaughey, W. P., Cohen, R. D. H., 1997. Effect of grazing system, stocking rate and season of use on diet quality and herbage availability of alfalfa–grass pastures. Can. J. Anim. Sci.77, 111–118.

Schönbach, P., Wan, H. W., Schiborra, A., Gierus, M., Bai, Y. F., Müller, K., Glindemann, T., Wang, C. J., Susenbeth, A., Taube, F., 2009. Short–term management and stocking rate effects of grazing sheep on herbage quality and productivity of Inner Mongolia steppe. Crop Pasture Sci.60, 963–974.

Schönbach, P., Wan, H. W., Gierus, M., Bai, Y. F., Müller, K., Lin, L. J., Susenbeth, A., Taube, F., 2011. Grassland responses to grazing: effects of grazing intensity and management system in an Inner Mongolian steppe ecosystem. Plant Soil 340, 103–115.

Sharrow, S. H., 1983. Forage standing crop and animal diets under rotational vs. continuous grazing. J. Range Manage.36, 447–450.

Stuth, J. W., Grose, P. S., Roath, L. R., 1987. Grazing dynamics of cattle stocked at heavy rates in a continuous and rotational grazed system. Appl. Anim. Behav. Sci.19, 1–9.

Su, Y. Z., Li, Y. L., Cui, H. Y., Zhao, W. Z., 2005. Influences of continuous grazing and livestock exclusion on soil properties in a degraded sandy grassland, Inner Mongolia, northern China. CATENA 59, 267–278.

Teague, W. R., Dowhower, S. L., 2003. Patch dynamics under rotational and continuous grazing management in large, heterogeneous paddocks. J. Arid Environ.53, 211–229.

Vallentine, J. F., 2001. Grazing managment. 2nd ed. San Diego, CA, USA: Academic Press.

Van Poollen, H. W., Lacey J, R., 1979. Herbage response to grazing systems and stocking intensities. J. Range Manage.32, 250–253.

Virgona, J. M., Avery, A. L., Graham, J. F., Orchard, B. A., 2000. Effects of grazing management on phalaris herbage mass and persistence in summer–dry environments. Aust. J. Exp. Agric.40, 171–184.

Walker, J. W., Heitschmidt, R. K., Demoraes, E. A., Kothmann, M. M., Dowhower, S. L., 1989. Quality and botanical composition of cattle diets under rotational and continuous grazing treatments. J. Range Manage.42, 239–242.

Wang, C. J., Tas, B. M., Glindemann, T., Müller, K., Schiborra, A., Schönbach, P., Gierus, M., Taube, F., Susenbeth, A., 2009a. Rotational and continuous grazing of sheep in the Inner Mongolian steppe of China. J. Anim. Physiol. Anim. Nutr.93, 245–252.

Wang, C. J., Tas, B. M., Glindemann, T., Rave, G., Schmidt, L., Weissbach, F., Susenbeth, A., 2009b. Fecal crude protein content as an estimate for the digestibility of forage in grazing sheep. Anim. Feed. Sci. Technol. 149, 199–208.

Wang, S. P., Wang, Y. F., Chen, Z. Z., 2001. The biological economical principle on sustainable development of grassland livestock in Inner Mongolia steppe. Acta Ecol. Sin.21, 617–623 (in Chinese).

Warner, J. R., Sharrow, S. H., 1984. Set stocking, rotational grazing and forward rotational grazing by sheep on western oregon hill pastures. Grass Forage Sci.39, 331–338.

Chapter 3 Effect of grazing system on feed intake and performance of sheep in a semi-arid grassland steppe

3 Effect of grazing system on feed intake and performance of sheep in a semi-arid grassland steppe

3.1 Abstract

Many studies evaluated the effects of different grazing management systems (GS) on biomass production and the nutritional quality of the rangeland vegetation in semi-arid regions of the world. However, less work has been done regarding their effects on diet digestibility, feed intake, and performance of grazing livestock. We therefore analyzed digestibility of ingested organic matter (dOM), organic matter intake (OMI), and live weight gain (LWG) of sheep grazing at two different GS's and determined whether GS effects may vary between grazing intensities (GI) due to differences in the amount and quality of herbage on offer. A grazing experiment was established in the Inner Mongolian steppe of China (E116°42′, N43°38′) in spring 2005. Two GS's were tested at 6 different GI's from very light to very heavy grazing. While in the alternating grazing system, grazing and hay-making were alternated annually between 2 adjacent plots, sheep grazed the same plots every year in the continuous grazing system. In July, August, and September 2009 and 2010, 4 sheep per plot were selected to determine feces excretion on 5 d per month using the external marker titanium dioxide, while dOM was estimated from fecal crude protein concentrations. OMI was calculated and sheep were weighed at the beginning of each month to determine their LWG.

GS did not affect dOM ($P=0.101$), OMI ($P=0.381$), and LWG of sheep ($P=0.701$). Across both GS's LWG decreased from 98 g/d at GI1 to 62 g/d at GI6 ($P<0.001$; $R^2=0.42$). Nevertheless, there were no significant interactions between GS and GI for all measured parameters ($P \geqslant 0.061$), indicating that differences between GS's were similar at all GI's and that alternating grazing was not able to compensate for the negative effects of very high GI on animal performance. In summary, our study showed that GI alternating grazing does not increase dOM, OMI, and hence, LWG of sheep in this environment even after four years of grazing. Nevertheless, in the long term, improvements in the productivity and feeding value of the steppe vegetation by an alternating pastoral grassland use might enhance revenues and ecological sustainability when compared to the common practice of continuous grazing at very high stocking rates.

3.2 Introduction

Grazing system (GS) is considered an important management tool that can maintain or even increase long-term rangeland and livestock production in pastoral farming systems (Long, 1986). In addition to the continuous (CON) grazing system in which a particular area is used for livestock grazing every year, several improved GS's were conceptualized. Their main aims are (1) to maintain rangelands in a productive state and (2) to make the most effective use of natural feed resources to generate income from animal products (Clark, 1994). In the alternating (ALT) grazing system hay-making and animal grazing are regularly alternated between two or more paddocks or areas (Merrill, 1954). This enables the vegetation to recover from grazing during hay-making years and returns organic matter and nutrients to the rangelands in grazing years through the disposal of animal feces and urine (Owens et al., 1989). ALT grazing might thus provide higher quantity and quality of herbage to grazing livestock, and could therefore increase the animals' nutrient and energy intakes, and thus their performance level (Heady, 1961).

Many studies analyzed the effects of ALT and CON on herbage production and quality. Reardon and Merrill (1976) stated that forage yields and litter accumulation were higher at ALT than at CON. Similarly, Clarke et al. (1943) showed in an earlier study that ALT grazing increased herbage production and was more favorable for the seasonal development and life cycle of the main forage species than CON grazing. Within the same research project as the presented study, Schönbach et al. (2011) determined a higher above-ground net primary production (ANPP) of the grassland steppe in Inner

Mongolia at ALT than at CON at heavy grazing intensities (GI). Moreover, Wan et al. (2011) concluded that above-ground biomass of the herbage species preferably grazed by sheep was less affected by moderate to high GI at ALT than at CON. Instead, few studies were conducted to clarify the effects of ALT on grazing livestock, and most of them focused on rather measuring animal performance than feed intake and quality. The objectives of our experiment were therefore to investigate the effects of ALT versus CON grazing on diet digestibility, feed intake, and live weight gain (LWG) of grazing sheep. We aimed to determine whether it can compensate for the negative effects of increasing GI and advancing plant maturation with proceeding vegetation period on these parameters. It was hypothesized that ALT increases energy intake and hence, animal performance by a higher mass and nutritional quality of the herbage on offer.

3.3 Materials and methods

3.3.1 Study area

The study site (116°42′E, 43°38′N) is located in the Xilin River Basin, Inner Mongolia Autonomous Region of China. The average altitude is approximately 1200 m above sea level. The semi-arid continental temperate steppe climate is characterized by a mean annual precipitation of 335 mm and a mean annual air temperature of 0.9℃ (weather data were collected at a station located close to our experimental areas in 1982—2008). Rainfall mainly occurs from May to August, and the vegetation period lasts for approximately 150 d from April to September (Bai et al., 2004). Monthly precipitations and mean monthly temperatures in 2009 and 2010 are shown in Figure 3.1. The dominant soil type is a chestnut soil (Chen and Wang, 2000). The steppe vegetation is dominated by the perennial rhizome grass *Leymus chinensis* Trin. and the perennial bunchgrass *Stipa grandis* P. A. Smirn (Bai et al., 2004). Annual yield of ANPP is 140 g dry matter (DM) /m^2 (Schönbach et al., 2011).

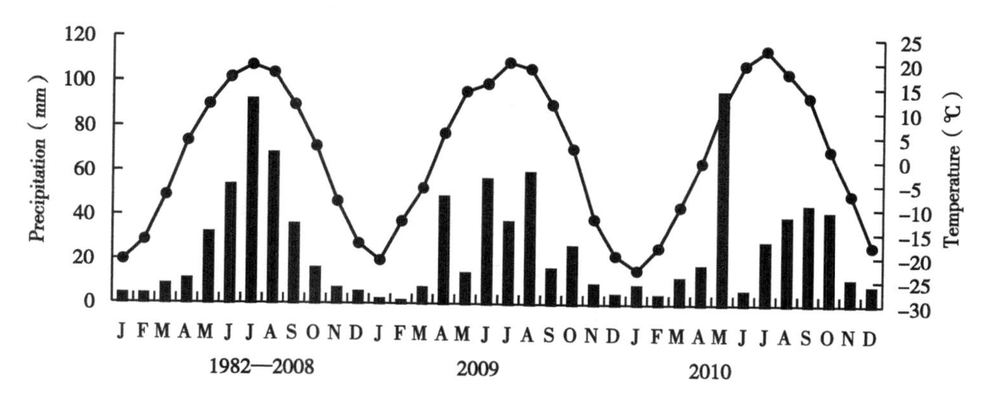

Figure 3.1 Monthly precipitation (mm, bars, primary y-axis) and mean monthly air temperature (℃, line and points, secondary y-axis) measured near the experimental area across 1982—2008 and in 2009 and 2010.

Livestock production has been and is still the main agricultural activity in Inner Mongolia. In the past, the land was extensively used for nomadic sheep production. Since 1950, the human population of the Xilingol league (202 580 km^2) rapidly increased from 200 000 in 1949 to approximately 950 000 in 2000 (Jiang et al., 2006). Correspondingly, the number of grazing animals increased by 18-fold from 1949 to 2000, so that the available grassland area decreased from 5 ha to 1 ha/sheep (Jiang et al., 2006). Moreover, the government strongly encouraged the nomadic families to settle down and abandon their traditional way of steppe use. Hence, grassland close to farms is currently used for intensive sheep and cattle grazing, while distant areas are only used for hay-making (Christensen et al., 2003). The constant removal of herbage on grazing areas reduces vegetation cover and hence, increases the risk of soil erosion. On hay-making areas, the lack of dung disposal and thus nutrient inputs to the grassland may negatively affect its long-term productivity (Owens and Shipitalo, 2009; Schönbach et al., 2009).

3.3.2 Experimental design

Two grazing experiments were conducted from mid June to mid September in 2009 and 2010 lasting for 92 d and 91 d, respectively. Outside of the grazing seasons, the grassland remained ungrazed. Two GS's were tested that had already been established in June 2005. For the ALT system, grazing and hay-making were alternated annually between two adjacent plots, whereas sheep grazed the same plots every year in the CON system. In each GS, 6 different GI treatments from very light (GI1), light (GI2), light-moderate (GI3), and moderate (GI4) to heavy (GI5) and very heavy (GI6) grazing were established. GI was defined by herbage allowance (HA) which will better describe GI than merely SR's, if herbage mass (HM) on offer strongly differs between experimental plots, years, or throughout the vegetation period (Sollenberger et al., 2005). HA target ranges were >12.0, >6.0 ~ 12.0, >4.5 ~ 6.0, >3.0 ~ 4.5, >1.5 ~ 3.0 and ≤1.5 kg DM/kg live weight (LW) for GI1 to GI6, respectively. The numbers of sheep per plot were adjusted to HM on offer at the middle of June, July, and August in order to maintain similar HA's across the grazing seasons and in both study years. Details of the experimental scheme are given in Table 3.1. Each treatment was replicated on 2 plots, a flat and a moderately sloped plot. Hence, measurements were carried out on a total of 24 plots (2 GS's · 6 GI's · 2 plots). Each plot had a size of 2 ha, except the GI1 plots which covered 4 ha to be able to keep a minimum of 8 sheep per plot.

3.3.3 Animals live weight gain

In 2009 and 2010, 315 and 337 female sheep of the Inner Mongolia fat-tailed breed were purchased from 2 local farmers. In both study years, the animals were about 15 months old when they were obtained, and neither pregnant nor lactating. Sheep were treated for internal parasites and sheared before the experiments started. During the grazing seasons, they had free access to water and mineral in lick stones and were allowed to graze day and night. Immediately after purchase, all sheep were weighed on two consecutive days using a portable electronic platform balance (0.1 kg accuracy). Subsequently, they were divided into four different LW groups (light < 30 kg LW, moderate >30~35kg LW, heavy >35~40kg LW, and very heavy >40 kg LW). Out of each LW group, sheep were randomly allocated to the grazing plots to equalize mean LW per plot. After 10 days of adaptation, sheep were weighed again to determine their initial LW in the middle of June. Across all GI's and GS's, it averaged (34.8±4.0) kg and (30.2±1.9) kg LW in June 2009 and 2010, respectively.

All animals were weighed again on the 11th and 12th of July, August, and September. The mean LW of the two days was used to calculate their daily LWG during each month according to the following equation:

$$LWG\ [g/d] = (LW_m\ [kg] - LW_{m-1}\ [kg]) / D_m \cdot 1\ 000 \qquad (6)$$

Where LW is the mean LW of the two weighings per month and D the number of days between weighing dates. m and m − 1 indicate two consecutive months (June, July, August, and September). Daily LWG per hectare (LWGh, g/ha) was calculated by multiplying the mean LWG of individual sheep (LWGs, g/d) by the SR (sheep/ha) in the respective plot.

Table 3.1 Mean live weight (LW) of sheep in the beginning of grazing season, stocking rates (SR), herbage allowances (HA) as well as herbage mass (HM) and chemical composition on the experimental plots during the grazing seasons in 2009 and 2010 (Least squares means; n=12)

Parameter	GS	Grazing intensity						SE	Mean	SEM
		1	2	3	4	5	6			
LW, kg	ALT	32. 8	32. 2	32. 8	32. 4	32. 4	32. 6	0. 1	32. 5	0. 1
	CON	32. 5	32. 6	32. 7	33. 0	32. 5	32. 4	0. 1	32. 6	0. 1
	Mean	32. 6	32. 4	32. 7	32. 7	32. 4	32. 5	0. 1	32. 5	0. 1
SR, sheep/ha	ALT	1. 9	3. 9	4. 8	6. 3	8. 4	10. 1	0. 3	5. 9	0. 2
	CON	2. 4	4. 4	5. 9	7. 0	8. 9	10. 4	0. 3	6. 5	0. 2
	Mean	2. 2	4. 2	5. 4	6. 7	8. 7	10. 3	0. 2	6. 2	0. 2

(attached table)

Parameter	GS	Grazing intensity						SE	Mean	SEM
		1	2	3	4	5	6			
HA, kg DM/kg LW	ALT	15.6	6.8	4.8	3.9	2.0	1.8	2.3	5.8	1.5
	CON	22.1	10.8	6.6	3.2	2.4	1.1	3.5	7.7	2.3
	Mean	18.9	8.8	5.7	3.6	2.2	1.5	2.1	6.8	1.4
HM, kg DM/ha	ALT	1 144	968	882	916	662	660	39	872	22
	CON	2 103	1 801	1 443	893	752	426	93	1 236	77
	Mean	1 624	1 384	1 162	905	707	543	52	1 054	48
CP, % DM	ALT	10.2	10.4	11.0	10.9	10.7	10.8	0.2	10.6	0.1
	CON	8.9	8.7	9.4	9.7	11.6	13.4	0.3	10.3	0.2
	Mean	9.5	9.6	10.2	10.3	11.2	12.1	0.2	10.5	0.1
NDF, % DM	ALT	69.7	69.7	69.4	69.8	70.0	69.4	0.2	69.7	0.1
	CON	70.8	71.2	70.8	70.7	69.2	67.5	0.2	70.1	0.2
	Mean	70.3	70.4	70.1	70.3	69.6	68.5	0.2	69.9	0.1
ADF, % DM	ALT	33.8	32.6	32.9	32.9	33.7	33.8	0.2	33.3	0.1
	CON	35.3	34.6	33.7	34.4	31.9	31.2	0.3	33.5	0.2
	Mean	34.5	33.6	33.3	33.7	32.8	32.5	0.2	33.4	0.1
ADL, % DM	ALT	4.8	4.8	4.8	4.9	5.0	4.9	0.1	4.9	0.1
	CON	5.1	5.0	4.8	5.1	4.7	4.6	0.1	4.9	0.1
	Mean	4.9	4.9	4.8	5.0	4.9	4.8	0.1	4.9	0.1
CDOM, % OM	ALT	63.8	63.8	64.0	63.3	62.8	63.3	0.3	63.5	0.1
	CON	61.2	60.9	61.9	61.2	64.4	66.0	0.4	62.6	0.3
	Mean	62.5	62.4	62.9	62.3	63.6	64.7	0.2	63.1	0.1

ADF, acid detergent fiber; ADL, acid detergent lignin; ALT, alternating grazing; CDOM, cellulase digestible organic matter; CON, continuous grazing; CP, crude protein; DM, dry matter; GS, grazing system; NDF, neutral detergent fiber; OM, organic matter; SE, standard error; SEM, standard error of mean.

For further descriptions of the GS and grazing intensity treatments see text.

3.3.4 Herbage allowance

Standing herbaceous biomass was determined by cutting the sward at 1 cm above ground level within a 2.0 m×0.25 m-frame in three representative areas per plot. The collected herbage material was oven-dried for 24 h at 60℃ and weighed to determine its DM content. Subsequently, HA's were calculated according to the formula given by Sollenberger et al. (2005):

$$HA\ [kg\ DM/kg\ LW] = (SB_1\ [kg\ DM]\ /LW_1\ [kg] + SB_2\ [kg\ DM]\ /LW_2\ [kg])\ /2 \quad (7)$$

Where SB is the standing herbaceous biomass per plot, DM is the dry matter concentration in herbage, and LW is the total live weight of all sheep per plot. Indices 1 and 2 represent two consecutive sampling days at the beginning of June, July, August, and September, respectively.

3.3.5 Digestibility of ingested herbage and feed intake

At the beginning of the grazing seasons, 4 sheep per plot were randomly chosen to determine digestibility of ingested organic matter (dOM) and organic matter intake (OMI). Daily OMI of sheep was calculated from dOM and daily fecal organic matter (FOM) excretion. FOM excretion was determined using the external marker titanium dioxide (TiO_2) assuming a fecal recovery rate of 100% (Glindemann et al., 2009a). For this, one gelatin capsule filled with 2.5 g TiO_2 (electronic balance with 0.001 g accuracy) was orally administered to the 4 sheep per plot each day for the first 10 days of July, August, and September, respectively. Immediately after marker application, fecal samples (approximately 25 g fresh matter per sheep) were taken from the rectum on days 6~10 (sampling period) and frozen. At the end of each sampling period, all samples were thawed and pooled to one sample per sheep (100 g fresh matter). The pooled samples were divided into two sub-samples. One sub-sample was oven-dried at 60℃ for 48 h, milled by a mixer, and used to determine fecal TiO_2 concentration according to procedures described by Glindemannet al. (2009a). The other sub-sample was analyzed for DM, crude ash, and crude protein (CP) concentrations. DM, crude ash, and nitrogen (N) were determined according to the methods of the Chinese Technical Committee for Feed Industry Standardization and the Chinese Association of Feed

Industry (2000).

dOM was estimated from CP ($=N\times6.25$) in FOM according to the non-linear regression equation given by Wang et al. (2009b):

$$dOM\ [\%] = 89.9-64.4\times\exp(-0.5774\times CP\ [g/kg\ FOM]/100) \quad (8)$$

Subsequently OMI of sheep was calculated using the following equation:

$$OMI\ [g/d] = FOM\ [g/d] / (100-dOM\ [\%]) \quad (9)$$

Intakes of digestible organic matter (DOMI) and metabolizable energy (MEI) were calculated by multiplying OMI by dOM or dietary metabolizable energy (ME) concentrations, respectively. The latter was estimated from dOM according to the formula of Aiple al. (1992):

$$ME\ [MJ/kg\ organic\ matter] = -0.9 + 0.170\times dOM\ [\%] \quad (10)$$

3.3.6 Statistical analysis

In total, 144 observations (2 GS's×6 GI's×2 plots×2 years×3 months) were obtained for all parameters. Data were analyzed using the Mixed Model procedure of SAS version 9.2 (SAS Institute Incorporated, Cary, NC, USA) to test for the effects of GS, GI, year, month, and their interactions. The following model was used:

$$y_{ijkl} = \mu + GS_i + GI_j + YE_k + M_l + GS\times GI_{ij} + GS\times YE_{ik} + GS\times M_{il} + GI\times YE_{jk} + GI\times M_{jl} + YE\times M_{kl} + GS\times GI\times YE_{ijk} + GS\times YE\times M_{ikl} + GI\times YE\times M_{jkl} + e_{ijkl}$$

Where μ is the overall mean, GS_i the grazing system (i = ALT and CON), GI_j the grazing intensities (j = 1, 2, 3, 4, 5, and 6), YE_k the year (k = 2009 and 2010), M_l the month (l=July, August, and September), and e_{ijkl} the random experimental error. All factors and their interactions were treated as fixed effects. Year was treated as repeated measurement. An autoregressive co-variance structure was chosen. When effects were significant ($P \leq 0.05$), the Tukey test was used for pair wise comparisons of least squares means. Linear or quadratic regression analyses were applied to analyze the effects of SR (independent variable) on dOM, OMI, LWGs, and LWGh (dependent variables). For this, means for each plot across the two study years were calculated and treated as one observation resulting in a total number of 24 observations for each parameter. Moreover, regression analysis were performed separately for each year using the mean values across the three months (n=12).

3.4 Results

3.4.1 Effects of grazing system and grazing intensity

The effects of GS and GI on measured parameters are presented in Table 3.2. GS did not affect dOM ($P=0.101$), OMI ($P=0.381$), DOMI ($P=0.209$), and MEI ($P=0.195$) as well as LWGs ($P=0.701$) and LWGh ($P=0.390$).

Mean dOM across both GS's, both years, and all months ranged from 0.546 at GI4 to 0.562 at GI6 ($P=0.001$). It differed between individual GI treatments ($P=0.010$) and according to the quadratic regression equation, first declined then increased with SR ($P=0.007$, $R^2=0.31$; Table 3.3). Moreover, no differences between GI's were determined for OMI ($P=0.327$), DOMI ($P=0.266$), and MEI ($P=0.258$). Although daily LWGs linearly decreased ($P<0.001$, $R^2=0.42$), daily LWGh increased with SR ($P<0.001$, $R^2=0.80$). However, the decrease in LWGs was only pronounced in 2009 ($P=0.011$, $R^2=0.44$ for ALT; $P=0.005$, $R^2=0.52$ for CON), whereas in 2010, LWGs was unaffected by SR ($P=0.731$, $R^2=-0.09$ for ALT; $P=0.041$, $R^2=0.29$ for CON; Figure 3.2).

Table 3.2 Effect of alternating (ALT) and continuous (CON) grazing on digestibility of ingested organic matter (dOM), daily intakes of organic matter (OMI), digestible organic matter (DOMI), and metabolizable energy (MEI), as well as daily live weight gain (LWG) of individual sheep (LWGs) and per hectare (LWGh) at different grazing intensities (GI) in 2009 and 2010 (Least squares means; n=12).

Parameter	GS	GI						SE	Mean	SEM	*P*-value*		
		1	2	3	4	5	6				GS	GI	GS×GI
SR, sheep/ha	ALT	1.9	3.9	4.8	6.3	8.5	10.2	0.1	6.0				
	CON	2.4	4.4	5.8	7.0	8.9	10.2	0.2	6.4				
	Mean	2.2	4.2	5.3	6.6	8.7	10.2	0.2	6.2	0.2			
HA, kg DM/kg LW	ALT	15.6	6.8	4.8	3.9	2.0	1.8	2.3	5.8				
	CON	22.1	10.8	6.6	3.2	2.4	1.1	3.5	7.7				
	Mean	18.9	8.8	5.7	3.6	2.2	1.5	2.1	6.8	1.4			
dOM	ALT	0.563	0.550	0.558	0.549	0.555	0.555	0.003	0.555				
	CON	0.552[ab]	0.544[a]	0.544[a]	0.543[a]	0.551[ab]	0.569[b]	0.003	0.551				
	Mean	0.557[ab]	0.547[a]	0.551[ab]	0.546[a]	0.553[ab]	0.562[b]	0.002	0.553	0.006	0.101	0.010	0.061
OMI, g/kg$^{0.75}$ LW	ALT	74.9	73.1	71.9	76.2	70.4	79.8	1.5	74.4				
	CON	76.7	71.3	72.7	76.1	72.7	68.7	1.3	73.1				
	Mean	75.8	72.2	72.3	76.1	71.5	74.2	1.0	73.7	0.36	0.381	0.327	0.112
DOMI, g/kg$^{0.75}$ LW	ALT	42.4	40.5	40.4	42.1	39.2	44.3	0.98	41.5				
	CON	42.4	39.1	39.7	41.7	40.2	39.2	0.87	40.4				
	Mean	42.4	39.8	40.0	41.9	39.7	41.8	0.66	40.9	0.21	0.209	0.266	0.405

(attached table)

Parameter	GS	GI						SE	Mean	SEM	*P*-value*		
		1	2	3	4	5	6				GS	GI	GS×GI
MEI, $MJ/kg^{0.75}$ LW	ALT	0. 65	0. 62	0. 62	0. 64	0. 60	0. 68	0. 02	0. 64				
	CON	0. 65	0. 60	0. 61	0. 64	0. 62	0. 60	0. 01	0. 62				
	Mean	0. 65	0. 61	0. 61	0. 64	0. 61	0. 64	0. 01	0. 63	0. 03	0. 195	0. 258	0. 456
LWGs, g	ALT	105	79	87	92	73	69	6	84				
	CON	91	94	92	87	72	55	5	82				
	Mean	98^{b}	86^{b}	89^{b}	89^{b}	72^{ab}	62^{a}	4	83	13	0. 701	0. 003	0. 501
LWGh, g/ha	ALT	199^{a}	305^{ab}	416^{abc}	573^{bc}	595^{bc}	678^{c}	44	461				
	CON	223^{a}	410^{ab}	515^{ab}	630^{b}	623^{b}	570^{b}	38	495				
	Mean	211^{a}	357^{ab}	465^{bc}	601^{c}	609^{c}	624^{c}	29	478	96	0. 390	< 0. 001	0. 660

[a,b,c] Within rows means without a common superscript differ at $P \leq 0.05$.

* Effects in bold characters were significant at $P \leq 0.05$.

There were no significant differences ($P > 0.05$) between ALT and CON at the same GI for any of the measured parameters.

GS, grazing system; HA, herbage allowance; LW, live weight;; SE, standard error; SEM, standard error of mean; SR, stocking rate.

Table 3.3 Parameters of the linear and non-linear regressions between stocking rate (SR; sheep/ha) and the digestibility of ingested organic matter (dOM), daily intakes of organic matter (OMI) and metabolizable energy (MEI), as well as the daily live weight gain (LWG) of individual sheep (LWGs) and per hectare (LWGh) at an alternating (ALT) and continuous (CON) grazing system in 2009 and 2010.

Parameter	Grazing system	Type of regression	a	b	c	*P*-value *	adjusted R^2
dOM	ALT	Linear	—	-0.001 (0.001)	0.559 (0.006)	0.446	-0.04
	CON	Quadratic	0.0005 (0.001)	-0.011 (0.002)	0.575 (0.007)	< 0.001	0.80
	Mean	Quadratic	0.0001 (0.001)	-0.009 (0.003)	0.575 (0.008)	0.007	0.31
OMI, g/kg$^{0.75}$ LW	ALT	Quadratic	0.0002 (0.001)	-0.003 (0.002)	0.079 (0.007)	0.384	0.01
	CON	Linear	—	-0.001 (0.001)	0.074 (0.002)	0.369	-0.01
	Mean	Linear	—	0.001 (0.001)	0.075 (0.002)	0.818	-0.05
MEI, MJ/kg$^{0.75}$ LW	ALT	Quadratic	0.0018 (0.002)	-0.029 (0.019)	0.701 (0.052)	0.297	0.07
	CON	Linear	—	-0.002 (0.004)	0.636 (0.029)	0.593	0.03
	Mean	Linear	—	-0.001 (0.003)	0.633 (0.019)	0.829	-0.04
LWGs, g	ALT	Linear	—	-3.54 (1.38)	104.6 (9.0)	0.029	0.33
	CON	Linear	—	-4.32 (1.37)	110.2 (9.8)	0.010	0.45
	Mean	Linear	—	-3.90 (0.93)	107.1 (6.3)	< 0.001	0.42
LWGh, g/ha	ALT	Linear	—	59.5 (7.4)	109.3 (48.2)	< 0.001	0.85
	CON	Linear	—	47.6 (11.6)	180.9 (82.4)	0.002	0.69
	Mean	Linear	—	53.7 (6.6)	142.3 (44.9)	< 0.001	0.74

* Probability values of the linear or quadratic relationships. Effects in bold characters were significant at $P \leq 0.05$.

Means for each plot across the two study years and three months were used. The number of observations was therefore n = 12 for each grazing system and n = 24 across both grazing systems. Regression equations are $y = ax + b$ for linear and $y = ax^2 + bx + c$ for quadratic regressions. Only parameters of the regressions with the highest adjusted R^2-values are given. Numbers inside parentheses represent one standard error.

LW, live weight.

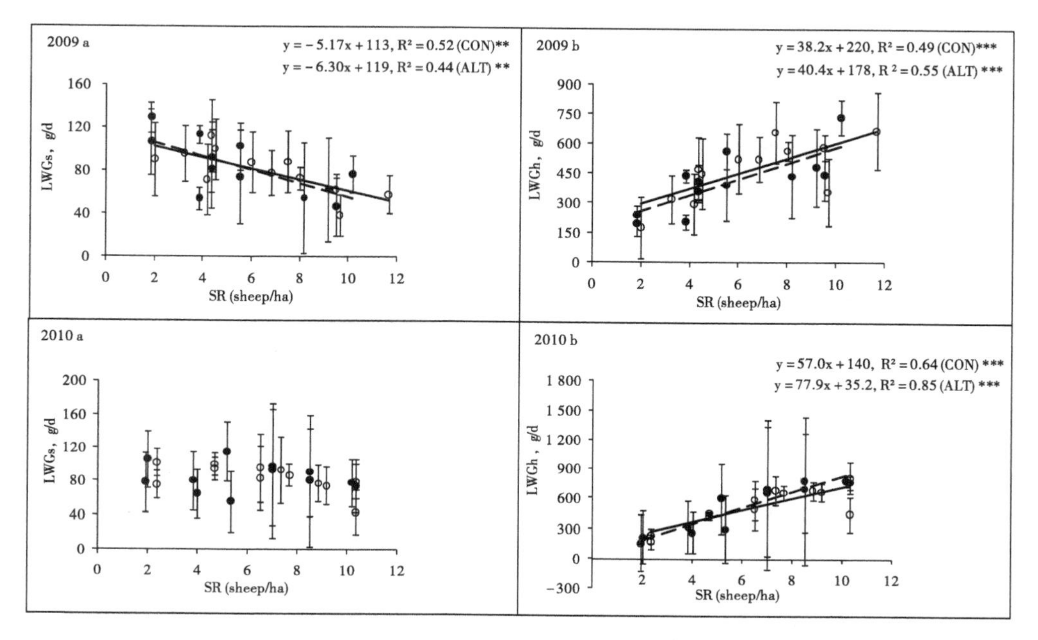

Figure 3.2 Linear relationships between stocking rate (SR; independent variable) and live weight gain (LWG) of individual sheep (LWGs, a; dependent variable) and per hectare (LWGh, b; dependent variable) in 2009 and 2010.

Circles (○) and points (●) represent the mean LWGs and LWGh at continuous (CON) and alternating (ALT) grazing, respectively, across the entire grazing seasons. Bars indicate standard errors from the means. "—" and "---" indicate the linear regression lines for CON and ALT, respectively. *, **, and *** highlight regressions that are significant at a level of $P \leqslant 0.05$, $P \leqslant 0.01$, and $P \leqslant 0.001$, respectively. The regression analysis is not significant in 2010a.

Interactions between GS and GI for feed intake ($P \geqslant 0.112$) and performance parameters did not occur ($P \geqslant 0.501$; Table 3.2). However, a tendency of an interaction between GS and GI was observed for dOM ($P=0.061$). While dOM was similar at all GI treatments at ALT ($P \geqslant 0.605$), it was higher at GI6 (0.569) than at GI2~GI4 (0.543~0.544; $P \leqslant 0.011$) at CON. No interactions between GS and month ($P \geqslant 0.617$) and between GI and month ($P \geqslant 0.173$) were found for any of the measured parameters, indicating that the effects of GS or GI did not differ between months and the changes in dOM, feed intake, and LWGs with proceeding grazing season were similar at both GS's and all GI treatments.

3.4.2 Effects of year and month

Across both GS's and all GI treatments, dOM and feed intake of sheep differed between years ($P < 0.001$ for all measured parameters; data not shown). In 2009, dOM, OMI, and MEI were 0.574, 79.7 g/kg$^{0.75}$ LW, and 0.71 MJ/kg$^{0.75}$ LW, respectively, and higher than in 2010 when corresponding values averaged 0.533 ($P< 0.001$), 67.6 g/kg$^{0.75}$ LW ($P< 0.001$), and 0.55 MJ/kg$^{0.75}$ LW ($P < 0.001$). However, at 81 g/d (2009) and 84 g/d (2010), no difference in LWGs was found between years ($P=0.600$). Daily LWGh was lower in 2009 (437 g/ha) than in the 2010 (519 g/ha; $P = 0.022$) due to slight differences in the mean SR.

All parameters were affected by month ($P< 0.001$ for all measured parameters; Table 3.4). dOM, OMI, and MEI were higher in July than in September ($P< 0.001$ for all three parameters). Correspondingly, a higher daily LWGs and LWGh were found in July (115 g, 678 g/ha) than in September (46 g, 230 g/ha; $P< 0.001$ for both parameters).

With the exception of LWGs ($P =0.474$) and LWGh ($P=0.178$), the interactions between year and month influenced all measured parameters ($P<0.001$ for dOM, OMI, and MEI; data not shown). While dOM ($P \leqslant 0.002$), OMI ($P< 0.001$), and MEI ($P< 0.001$) were higher in July than in August and September in 2009, dOM and feed and energy intakes were higher in July and August than in September in 2010 ($P< 0.001$ for all three parameters; data not shown).

Table 3.4 Effect of alternating (ALT) and continuous (CON) grazing on digestibility of ingested organic matter (dOM), daily intakes of organic matter (OMI), digestible organic matter (DOMI), and metabolizable energy (MEI) as well as daily live weight gain (LWG) of individual sheep (LWGs) and per hectare (LWGh) during different months of grazing seasons in 2009 and 2010 (Least squares means; n=24).

Parameter	GS	Month						P-value *						
		Jul.	Aug.	Sept.	SE	Mean	SEM	M	YE	M×YE	GS×M	GS×YE	GI×YE	GI×M
SR, sheep/ha	ALT	5.8	5.9	6.0	0.3	6.0								
	CON	6.1	6.5	6.8	0.3	6.4								
	Mean	6.0	6.2	6.4	0.2	6.2	0.1							
HA, kg DM/kg LW	ALT	6.2	6.4	4.9	0.2	5.8								
	CON	9.0	8.1	5.9	0.2	7.7								
	Mean	7.6	7.3	5.4	0.2	6.8	0.2							
dOM	ALT	0.566[b]	0.554[a]	0.545[a]	0.002	0.555								
	CON	0.562[b]	0.547[a]	0.542[a]	0.002	0.551								
	Mean	0.564[b]	0.551[a]	0.544[a]	0.002	0.553	0.004	< 0.001	< 0.001	< 0.001	0.860	0.054	0.086	0.340
OMI, $g/kg^{0.75}$ LW	ALT	81.2[b]	74.7[b]	67.1[a]	1.5	74.3								
	CON	80.7[b]	71.1[a]	66.8[a]	1.3	73.0								
	Mean	80.9[c]	73.2[b]	66.9[a]	1.0	73.7	2.53	< 0.001	< 0.001	< 0.001	0.689	0.764	0.027	0.938
DOMI, $g/kg^{0.75}$ LW	ALT	46.2[c]	41.1[b]	36.8[a]	0.98	41.5								
	CON	45.5[b]	39.2[a]	36.4[a]	0.87	40.4								
	Mean	45.8[c]	40.3[b]	36.6[a]	0.66	40.9	1.48	< 0.001	< 0.001	< 0.001	0.674	0.454	0.015	0.862

(attached table)

Parameter	GS	Month						P-value *						
		Jul.	Aug.	Sept.	SE	Mean	SEM	M	YE	M×YE	GS×M	GS×YE	GI×YE	GI×M
MEI, MJ/kg$^{0.75}$ LW	ALT	0.71^{c}	0.64^{b}	0.57^{a}	0.02	0.64								
	CON	0.70^{b}	0.60^{a}	0.56^{a}	0.01	0.62								
	Mean	0.71^{c}	0.62^{b}	0.56^{a}	0.01	0.63	0.02	< 0.001	< 0.001	< 0.001	0.684	0.421	0.014	0.945
LWGs, g	ALT	114^{c}	87^{b}	50^{a}	6	84								
	CON	116^{c}	88^{b}	42^{a}	5	82								
	Mean	115^{c}	87^{b}	46^{a}	4	83	9	< 0.001	0.600	0.474	0.647	0.843	0.664	0.173
LWGh, g/ha	ALT	682^{b}	413^{b}	218^{a}	44	461								
	CON	674^{b}	569^{b}	242^{a}	38	495								
	Mean	678^{c}	526^{b}	230^{a}	29	478	48	< 0.001	0.010	0.178	0.617	0.416	0.249	< 0.001

[a,b,c] Within rows means without a common superscript differ at $P \leqslant 0.05$. There were no significant differences ($P \leqslant 0.05$) between ALT and CON.

* Effects in bold characters were significant at $P \leqslant 0.05$. None of the effects of the triple interactions between GS, GI, YE, and M weresignificant at $P \leqslant 0.05$ and were thus not included in the table. Aug., August; GI, grazing intensity; GS, grazing system; HA, herbage allowance; Jul., July; LW, live weight; M, month;; SE, standard error; SEM, standard error of mean; Sept., September; SR, stocking rate; YE, year.

3.5 Discussion

3.5.1 Effects of grazing system

Previous studies within the frame of the same research project analyzed the effects of GS on biomass production and plant community structure of the grassland vegetation as well as the nutritional quality of herbage. Based on data collected during 2007 and 2008, Schönbach et al. (2011) found a higher ANPP, soil coverage, litter accumulation of the steppe vegetation at ALT grazing (defined as "mixed grazing" in their study), in particular at very high GI. However, the number of sheep was adjusted to HM on offer to maintain similar HA's between GS's, study years, and across the grazing seasons. Moreover, GS did not affect chemical composition of the pasture plants in 2005 and 2006 (Schönbach et al., 2009) and herbage CP, ME, neutral detergent fiber, and acid detergent fiber concentrations were also very similar at both GS's in the present study (Table 3.1). Altogether, this might explain why no differences in dOM, OMI, and hence, MEI and LWGs were found between ALT and CON sheep. To our knowledge, little information is available in literature on the effect of GS on dOM and OMI of grazing ruminants. The lack of a GS effect on LWGs confirms results of Lin et al. (2012) who analyzed the performance of sheep at both GS's within the frame of the same research project during 2005—2009. The authors suggested that LWGs was similar at ALT and CON because of similar digestibility and nutrient concentrations of the herbage on offer and the monthly adjustment of SR's at both GS's. Similarly, Allan (1997) found no differences between GS's in LWGs and wool growth of sheep grazing a tussock pasture in New Zealand (760 ~ 930m altitude, 500 mm annual rainfall). The author speculated that the duration of the experiment of only 6 years was too short to clearly establish GS effects on herbage and thus, animal performance. Moreover, the low and variable distribution in annual rainfall may greatly influence the GS effects on herbage parameters, and hence, feed intake and animal performance (Schönbach et al.,

2012). Hence, GS effects on the structure of the plant community were stronger in wet than in dry years in a study by Sternberg et al. (2000) who studied a Mediterranean rangeland vegetation in North-Eastern Israel. Altogether, this might explain why the expected advantage of ALT was not found in our study.

Across both study years and all GI's, HM on offer was 42% higher at CON than at ALT. However, this did not lead to a higher LWGh, since mean SR's were only slightly different between CON (6.5 sheep/ha± 0.3 sheep/ha) and ALT (5.9 sheep/ha± 0.3 sheep/ha). The lower HM on offer across all GI's at ALT was solely due to differences in HM at GI1-3. As a consequence, HA's at these GI's were higher at ALT than at CON. Nevertheless, they were within the defined HA classes (except GI3) so that the effects of the higher HA's at ALT on the measured parameters appear unlikely.

3.5.2 Interaction between grazing system and grazing intensity

No interactions between GS and GI were found for any of the measured animal parameters, indicating that even at very high GI with low HM on offer, ALT is not superior to CON. Results contradict with conclusions drawn in earlier studies within the frame of the same research project. Schönbach et al. (2012) found that LWGs clearly decreased with increasing SR at CON in 2005 and 2006. Instead, Müller et al. (2012) reported that LWGs of ALT sheep only decreased with increasing SR in 2005 when annual rainfall was very low, whereas in 2006 and 2007, it was similar at all GI's. The authors therefore speculated that ALT might have mitigated the negative effects of heavy GI on LWGs. Likewise, increasing SR strongly reduced ANPP at CON in 2005—2008, whereas grassland at ALT appeared to be more resilient to heavy grazing (Schönbach et al., 2011). Moreover, above-ground biomass of *L. chinensis*, one of the dominant grass species in the Inner Mongolian steppe that is preferably grazed by sheep (Wang, 2000), was lower at high than at moderate to low GI's at CON in July 2005—2009 (Wan et al., 2011). Instead GI did not affect above-ground biomass of this species at ALT (Wan et al., 2011). Hence, ANPP and botanical composition data underline the advantage of ALT grassland use.

Despite a strong decline in HA with increasing GI, there were only minor or no effects of GI on dOM, OMI, and MEI of sheep, neither at ALT nor at CON. Moreover, no significant relationships between SR and digestibility and intake parameters were determined according to the regression analyses for both GS's. Sheep can partly compensate for a decrease in herbage availability by increasing their grazing time and/or the frequency of their bites (Animut et al., 2005; Lin et al., 2011). Additionally, selective feeding behavior may enable sheep to maintain dOM at different HA's (Garcia et al., 2003). Altogether this explains why no differences between GI's in dOM and feed intake of sheep and thus, no interactions between GI and GS were found for these parameters.

Nevertheless, a strong decline in LWGs with increasing SR was observed at both GS's in 2009, indicating elevated ME requirements for grazing activities of sheep at higher GI's in this year (Animut et al., 2005; Lin et al., 2011). Such decreases in LWGs were not observed in 2010 or in most of the previous study years within the research project (Müller et al., 2012; Schönbach et al., 2012) which was related to differences in the amount and distribution of annual precipitation and hence, the quantity and quality of herbage re-growth during the grazing season (Bai et al., 2008; Schönbach et al., 2009). Despite the clear decline in LWGs with increasing GI in 2009, no effects of the interactions between GS and GI or GS, GI, and year were determined for LWGs and LWGh. Moreover, estimates of the intercept values and slopes of the linear regressions were almost identical at both GS's which not only contradicts with our hypothesis and the conclusions drawn by Müller et al. (2012) and Schönbach et al. (2012), but is also not in line with the positive effect of ALT observed in herbage parameters (Schönbach et al., 2011). The limited duration of the research project, the inter- and intra-annual variability in precipitation and ambient temperatures (see previous section), and only minor differences in herbage quantity and quality parameters between GS's in 2009 and 2010 (Ren et al., 2012; see Table 1) might explain why the ALT use of the grassland steppe neither improved feed intake and LWGs of sheep nor compensated for the negative effects of moderate to high SR's on animal performance. This underlines that changes in the rangeland vegetation do not necessarily increase LWG of grazing livestock. Furthermore, solely measuring animal performance will not capture (initial) positive effects on the grassland vegetation which might, nevertheless, enhance overall livestock production in the long term.

3.5.3 Effects of month

Due to advanced plant maturation, limited plant re-growth (Schönbach et al., 2009), and hence, a pronounced decline in the nutritional value of herbage on offer, dOM, OMI, and MEI decreased with proceeding grazing season. This is consistent with results of previous studies (Glindemann et al., 2009b; Wang et al., 2009a; Müller et al., 2012) in the same project and Garcia et al. (2003) who analyzed dOM and OMI of sheep in a semi-arid highland area in Central France. As a consequence, LWGs and LWGh strongly declined from July to September. In contrast to our expectations, no interactions between GS and month or GS, GI, and month were found. At such high dietary fiber concentrations (see Table 1), any further increases in NDF contents from July to September combined with decreases in fiber degradability and thus overall diet digestibility will strongly reduce OMI and LWG of sheep (Siebert and Hunter, 1977). Since HA's was kept constant across the grazing seasons, the decrease in herbage quality due to rapid plant maturation during the short vegetation period of only 150 d could outweigh any positive effects of GS on herbage and animal parameters. Nevertheless, long-term shifts in the botanical composition of herbage, increased ground cover and hence, soil water contents might not only enhance overall nutritional quality of the grassland vegetation, but also lower the rate of plant maturation (Ferraro and Oesterheld, 2002; Schönbach et al., 2011).

3.6 Conclusions

ALT grazing does not improve feed intake and LWG of sheep grazing semi-arid grassland steppes in a short term and cannot mitigate the negative effects of very high GI on animal performance. Besides study duration, low rainfall, and the variability in climatic conditions, the adaptive capacity of grazing livestock might explain the lack of direct effects. Nevertheless, ALT grazing enhanced production and ground coverage of herbage and its botanical composition in many studies. Hence, solely measuring either herbage or animal performance parameters is insufficient to comprehensively evaluate GS effects in pastoral ecosystems. Moreover, this suggests that an ALT use of grasslands in this and similar environments might increase revenues and ecological sustainability of livestock production in a long term when compared to the common practice of continuous grazing at high stocking rates.

3.7 References

Aiple, K. P., Steingass, H., Menke, K. H., 1992. Suitability of a buffered fecal suspension as the inoculum in the Hohenheim Gas Test.1. Modification of the method and its ability in the prediction of organic-matter digestibility and metabolizable energy content of ruminant feeds compared with rumen fluid as inoculum. J. Anim. Physiol. Anim. Nutr.67, 57-66.

Allan, B. E., 1997. Grazing management of oversown tussock country 3. Effects on liveweight and wool growth of Merino wethers. N. Z. J. Agric. Res. 40, 437-447.

Animut, G., Goetsch, A. L., Aiken, G. E., Puchala, R., Detweiler, G., Krehbiel, C. R., Merkel, R. C., Sahlu, T., Dawson, L. J., Johnson, Z. B., Gipson, T. A., 2005. Grazing behavior and energy expenditure by sheep and goats co-grazing grass/forb pastures at three stocking rates. Small Ruminant Res.59, 191-201.

Bai, Y., Han, X., Wu, J., Chen, Z., Li, L., 2004. Ecosystem stability and compensatory effects in the Inner Mongolia grassland. Nature 431, 181-184.

Bai, Y. F., Wu, J. G., Xing, Q., Pan, Q. M., Huang, J. H., Yang, D. L., Han, X. G., 2008. Primary production and rain use efficiency across a precipitation gradient on the Mongolia plateau. Ecology 89, 2140–2153.

Chen, Z. Z., Wang, S. P., 2000. Plant community structure, productivity, and development, in: Chen, Z. Z. and Wang, S. P. (Eds), Typical steppe Ecosystem of China. Science Press, Beijing, China. pp.9–45 (in Chinese).

Chinese Technical Committee for Feed Industry Standardization and the Chinese Association of Feed Industry, 2000. Chinese Technical Committee for Feed Industry Standardization and the Chinese Association of Feed Industry. China Standard Press, Beijing, China (in Chinese).

Christensen, L., Coughenour, M. B., Ellis, J. E., Chen, Z. Z., 2003. Sustainability of Inner Mongolian grasslands: Application of the Savanna model. J. Range Manage.56, 319–327.

Clark, D. A., 1994. Grazing for pasture management in New Zealand, in: David, R. K., David, L. M. (Eds.), Pasture management: Technology for the 21st Century. CSIRO, Canberra, pp.105–106.

Clarke, S. E., Tisdale, E. W., Skoglund, N. A., 1943. The effects of climate and grazing practices on short-grass prairie vegetation. C. Dept. Agr. Tech. Bull. 46, 53.

Ferraro, D. O., Oesterheld, M., 2002. Effect of defoliation on grass growth. A quantitative review. Oikos 98, 125–133.

Garcia, F., Baumont, R., Carrere, P., Soussana, J. F., 2003. The ability of sheep at different stocking rates to maintain the quality and quantity of their diet during the grazing season. J. Agric. Sci.140, 113–124.

Glindemann, T., Tas, B. M., Wang, C., Alvers, S., Susenbeth, A., 2009a. Evaluation of titanium dioxide as an inert marker for estimating faecal excretion in grazing sheep. Anim. Feed Sci. Technol.152, 186–197.

Glindemann, T., Wang, C., Tas, B. M., Schiborra, A., Gierus, M.,

Taube, F., Susenbeth, A., 2009b. Impact of grazing intensity on herbage intake, composition, and digestibility and on live weight gain of sheep on the Inner Mongolian steppe. Livest. Sci.124, 142-147.

Heady, H. F., 1961. Continuous vs. specialized grazing systems: a review and application to the California annual type. J. Range Manage.14, 182-193.

Jiang, G. M., Han, X. G., Wu, J. G., 2006. Restoration and management of the Inner Mongolia grassland require a sustainable strategy. Ambio 35, 269-270.

Lin, L. J., Dickhoefer, U., Müller, K., Wang, C. J., Glindemann, T., Hao, J., Wan, H. W., Schönbach, P., Gierus, M., Taube, F., Susenbeth, A., 2012. Growth of sheep as affected by grazing system and grazing intensity in the steppe of Inner Mongolia, China. Livest. Sci.144, 140-147.

Lin, L. J., Dickhoefer, U., Müller, K., Wurina, Susenbeth, A., 2011. Grazing behavior of sheep at different stocking rates in the Inner Mongolian steppe, China. Appl. Anim. Behav. Sci.129, 36-42.

Long, G. A., 1986. Management of Grazing System, in: Joss, P. J., Lynch, P. W., and Williams, O. B. (Eds.), Rangelands: A resource under siege. Cambridge University Press, Cambridge, UK, pp.206-211.

Merrill, L. B., 1954. A variation of deferred rotation grazing for use under southwest range conditions. J. Range Manage.7, 152-154.

Müller, K., Dickhoefer, U., Lin, L., Glindemann, T., Wang, C., Schönbach, P., Wan, H. W., Schiborra, A., Tas, B. M., Gierus, M., Taube, F., Susenbeth, A., 2014. Impact of grazing intensity on herbage quality, feed intake, and live weight gain of sheep grazing the steppe of Inner Mongolia, China. J. Agri. Sci.152 (1) 153-165.

Owens, L. B., Edwards, W. M., Vankeuren, R. W., 1989. Sediment and nutrient losses from an unimproved, all-year grazed watershed. J. Environ. Qual. 18, 232-238.

Owens, L. B., Shipitalo, M. J., 2009. Runoff quality evaluations of continuous and rotational over-wintering systems for beef cows. Agric. Ecosyst. Environ. 129, 482-490.

Reardon, P. O., Merrill, L. B., 1976. Vegetative response under various grazing management systems in the Edwards Plateau of Texas. J. Range Manage. 29, 195-198.

Ren, H. Y., W., W. H., Schönbach, P., Gierus, M., Taube, F., 2012. Forage nutritional characteristics and yield dynamics in a grazed typical steppe ecosystem of Inner Mongolia, China. Plant Soil (submitted).

Schönbach, P., Wan, H., Schiborra, A., Gierus, M., Bai, Y., Müller, K., Glindemann, T., Wang, C., Susenbeth, A., Taube, F., 2009. Short-term management and stocking rate effects of grazing sheep on herbage quality and productivity of Inner Mongolia steppe. Crop Pasture Sci. 60, 963-974.

Schönbach, P., Wan, H. W., Gierus, M., Bai, Y. F., Müller, K., Lin, L. J., Susenbeth, A., Taube, F., 2011. Grassland responses to grazing: effects of grazing intensity and management system in an Inner Mongolian steppe ecosystem. Plant Soil 340, 103-115.

Schönbach, P., Wan, H. W., Gierus, M., Loges, R., Müller, K., Lin, L. J., Susenbeth, A., Taube, F., 2012. Effects of grazing and precipitation on herbage production, herbage nutritive value and performance of sheep in continental steppe. Grass Forage Sci., in press, doi: 10.1111/j.1365-2494.2012.00874. x.

Siebert, B. D., Hunter, R. A., 1977. Prediction of herbage intake and live weight gain of cattle grazing tropical pastures from the composition of the diet. Agric. Sys.2, 199-208.

Sollenberger, L. E., Moore, J. E., Allen, V. G., Pedreira, C. G. S., 2005. Reporting forage allowance in grazing experiments. Crop Sci.45, 896-900.

Sternberg, M., Gutman, M., Perevolotsky, A., Ungar, E. D., Kigel, J., 2000. Vegetation response to grazing management in a Mediterranean herbaceous community: a functional group approach. J. Appl. Ecol.37, 224–237.

Wan, H. W., Bai, Y. F., Schonbach, P., Gierus, M., Taube, F., 2011. Effects of grazing management system on plant community structure and functioning in a semiarid steppe: scaling from species to community. Plant Soil 340, 215–226.

Wang, S. P., 2000. The dietary composition of fine wool sheep and plant diversity in Inner Mongolia steppe. Acta Ecol. Sin.20, 951–957 (in Chinese).

Wang, C. J., Tas, B. M., Glindemann, T., Müller, K., Schiborra, A., Schoenbach, P., Gierus, M., Taube, F., Susenbeth, A., 2009a. Rotational and continuous grazing of sheep in the Inner Mongolian steppe of China. J. Anim. Physiol. Anim. Nutr.93, 245–252.

Wang, C. J., Tas, B. M., Glindemann, T., Rave, G., Schmidt, L., Weissbach, F., Susenbeth, A., 2009b. Fecal crude protein content as an estimate for the digestibility of forage in grazing sheep. Anim. Feed Sci. Technol. 149, 199–208.

Chapter 4 General discussion

4 General discussion

4.1 Grazing management systems

Application of the improved and adapted management techniques may alter current livestock grazing systems and thereby ensure the sustainability of rangeland use (Bailey, 2004). Besides stocking rate (SR), supplementation, duration of grazing period, fertilization, and burning, grazing system (GS) is also considered as a main management tool that can enhance grassland production and animal performance. However, as mentioned in Chapter 1.4, results of previous studies about effects of GS on herbage and animal parameters are inconsistent (Table 4.1). The concern of the present thesis is the comparison of rotational (ROT; Chapter 2) and alternating (ALT; Chapter 3) grazing with continuous (CON) grazing, respectively. The study aims to identify the important factors which define a suitable grazing system to maintain or even increase long term animal productivity without reducing grassland productivity in the Inner Mongolian steppe. Digestibility of ingested herbage (dOM), feed intake, and live weight gain (LWG) of animals were measured in the present study. Combing the results of presented and previous studies that evaluated GS effects (Table 4.1), we conclude that SR, climate conditions, the animals' age and species, and the duration of practicing a specific GS might influence the extent of these effects.

According to results of Chapter 2, ROT is inferior to CON at a moderate SR (4.2 sheep/ha ± 0.3 sheep/ha) regarding herbage mass and chemical compositions of herbage on offer, as well as feed intake and LWG of individual sheep. SR is frequently discussed as an important factor determining the effects of GS on herbage quality and production, and thus animal performance (Hacker and Richmond, 1994; Simon et al., 1995; Schönbach et al., 2009; Schönbach et al., 2011). Due to the fact that relative growth rate of herbage production increases with SR (Schönbach et al., 2011),

Table4.1 Literature review of studies that compared different grazing systems with the continuous grazing system.

Author	Year	GS	Duration Year	Duration M	SR (animal/ha)	Annual rainfall (mm)	Species	Age month	Study site	Digestibility of ingested herbage (%)	OMI (g/sheep/d)	LWG (g/sheep/d)	Milk (kg/animal/d)	HM (DM/m^2)
Clarke et al.	1943	ALT, CON	—	—	—	~334	cattle	—	Alberta and Saskatchewan, Canada	—	—	—	—	ALT>CON
Campling et al.	1958	STR, CON	2	4	—	—	cow	—	—	—	—	—	STR=CON	—
Holmes and Osman	1960	STR, CON	1	4	—	—	cow	—	—	STR=CON	STR<CON	STR=CON	STR=CON	—
Reardon and Merrill	1976	ALT, CON	20	—	—	~700	sheep/cattle	—	Texas, USA	—	—	—	—	ALT>CON
Jamieson and Hodgson	1979	STR, ROT	3	1.5	—	—	calve	4~9	Alaska, USA	—	—	—	STR=CON	—
Warner and Sharrow	1984	ROT, CON	3	—	—	530	ewe	—	Oregon, USA	—	—	ROT=CON	—	ROT=CON
Hodgson	1985	TLG, CON	—	—	—	—	—	—	—	—	TLG<CON	TLG>CON	—	—
Holechek et al.	1987	ALT, CON	5	4	2.5	530	heifer	12	Oregon, USA	ALT=CON	—	ALT=CON	—	—
Grant et al.	1988	ALT, CON	—	—	—	—	sheep	—	—	—	—	—	—	ALT=CON
Heitschmidt et al.	1990	ALT, CON	4	12	5.3	~700	cow/calf	—	Texas, USA	—	—	ALT=CON	—	—
Dowling et al.	1996	ALT, CON	6	—	5~7	~600	sheep/cattle	—	New South Wales, Australia	—	—	—	—	ALT>CON

(attached table)

Author	Year	GS	Duration Year	M	SR (animal/ha)	Annual rainfall (mm)	Species	Age month	Study site	Digestibility of ingested herbage (%)	OMI (g/sheep/d)	LWG (g/sheep/d)	Milk (kg/animal/d)	HM (DM/m^2)
Hafley	1996	ROT, CON	2	3	3.0, 6.0	~1526	steer	12	Louisiana, USA	ROT<CON	—	ROT<CON	—	—
Allan	1997	ROT, ALT, CON	6	6	1.9, 3.0, 4.1	500	wether	12	New Zealand	—	—	ROT=ALT =CON	—	ROT>ALT =CON
Popp et al.	1997	ROT, CON	3	—	2.2, 1, 1	~ 450	cattle	—	Brandon, Canada	—	ROT=CON	—	—	ROT=CON
Virgona et al.	2000	ROT, CON	3	12	—	>600	sheep	—	New South Wales and Victoria, Australia	—	—	—	—	ROT>CON
Davies et al.	2001	ALT, CON	3	—	—	580	ewe	—	Bakers Hill, Western Australia	—	—	ALT=CON	—	ALT>CON
Lodge et al.	2003	ALT, CON	4	12	4.0, 6.0, 8.0	~689	wether	—	New South Wales, Australia	—	—	—	—	ALT>CON
Alvarez-Rodriguez et al.	2007	RAT, CON	1	3	—	—	ewe/lamb	—	Spain	—	—	RAT>CON	RAT=CON	—
Briske et al.	2008	ROT, CON	—	—	—	—	sheep/cattle	—	—	—	—	ROT≤CON	—	—
Derner et al.	2008	ROT, CON	16	4	0.2, 0.3, 0.4	381	steer	12	Wyoming, USA	—	—	ROT<CON	—	—
Joy et al.	2008	RAT, CON	1	4	—	—	ewe/lamb	—	Spain	—	—	—	RAT=CON	—
Schönbach et al.	2009	ALT, CON	2	3	1.5~9.0	330	ewe	15	Inner Mongolia, China	—	—	—	—	ALT=CON

(attached table)

Author	Year	GS	Duration Year	M	SR (animal/ha)	Annual rainfall (mm)	Species	Age month	Study site	Digestibility of ingested herbage (%)	OMI (g/sheep/d)	LWG (g/sheep/d)	Milk (kg/animal/d)	HM (DM/m^2)
Wang et al.	2009	ROT, CON	2	3	4.5	330	ewe	15	Inner Mongolia, China	ROT<CON	ROT<CON	ROT=CON	—	ROT=CON
Miller et al.	2010	ROT, CON	1	1.5	—	—	lamb	—	Tasmania, Australia	—	—	ROT=CON	—	—
Nie and Zollinger	2011	ALT, CON	5	12	—	451	sheep	—	Southern Australia	—	—	—	—	ALT>CON

ALT, alternating grazing; CON, continuous grazing; DM, dry matter; GS, grazing system; HM, herbage mass; LWG, live weight gain per animal; OMI, organic matter intake; ROT, rotational grazing; SR, stocking rate during the study period; STR, strip grazing; TLG, time limited grazing.

the positive effects of resting periods regarding herbage regrowth as well as animal production might be impaired when a low SR is used (Warner and Sharrow, 1984). This seems to be the reason why LWG of individual sheep was lower at ROT than at CON at relative lower SR in the present study. Similarly, Allan (1985, 1997) reported that herbage regrowth and animal performance increased with increasing SR at ROT. Warner and Sharrow (1984) stated that LWG of animal was higher at ROT than at CON plots at a higher SR, while they were identical at both systems at a lower SR (see Table 4.1).Information given by the publications were used to explain why ROT was identical or even inferior to CON in the first experiment with 4.2 sheep/ha. Likewise, no positive effect of ALT on feed intake and LWG at high SR were observed in the second experiment while above-ground net primary production (ANPP) was higher at ALT than at CON at a moderate to heavy SR (Schönbach et al., 2011). Therefore, it is concluded that the effect of SR on animal performance vary between different GS.

Besides SR, amount of rainfall is another important factor which influences the system effects. Many studies have shown the positive effects of rainfall on herbage production (Fynn and O'Connor, 2000; Patton et al., 2007; Bai et al., 2008; Schönbach et al., 2009) and herbage quality (Schönbach et al., 2009) on semi-arid grasslands. However, Heitschmidt et al. (1987) reported that herbage quality in the rest areas at ROT might decrease at a higher rainfall than the average of 682 mm due to the lack of continuous forage removal. This might have led to a high proportion of mature and dead herbage (Heitschmidt et al., 1987). Furthermore, herbage regrowth is limited by a low rainfall. Therefore, it is evident that the positive effects of ROT on herbage quality and thus, animal feed intake and LWG might vary between years due to the different amounts of precipitation. The results were consistent with our finding that metabolizable energy content of herbage and LWG of sheep were lower at ROT than at CON in the last study year at a relatively high annual rainfall (389 mm; average rainfall of other three years was 266 mm), while LWG was identical at both GS in the previous 3 study years. In the last year, higher precipitation increased regrowth of herbage by grazing and thus was less mature in CON plots. Hence, herbage quality was lower at ROT than at CON, which caused a lower LWG at ROT. Similarly, lower LWG of cattle was observed at ROT than at CON in Louisiana, USA at annual rainfall of 1526 mm (Halfey, 1996). Hence, level of rainfall is an important factor for herbage production but not the sole determinant for dOM, feed intake, and LWG of animals.

According to Manley et al. (1997), longer time periods are necessary to indicate plant botanical composition changes caused by grazing on rangelands. However, most of the studies evaluating that herbage quality and quantity and performance of grazing animals at different GS's were conducted for a relatively short term of three to five years (see Table 4.1). Perhaps the positive effects of specialized GS's may become manifest after long term grazing (Derner et al., 2008). Hence, Reardon and Merrill (1976) reported that ALT increased the herbage production after 20 years of grazing sheep and cattle, which is probably due to the recovery phases for grazed swards and the seed production during un-grazed years (Derner et al., 2008). This argument was used to explain why no differences in LWG (Allan, 1997) and herbage mass (Schönbach et al., 2009) were found between ALT and CON in the 6-year and the 2-year study, respectively. Similarly, no positive effects of ALT on animal LWG in the present study (Chapter 3) might be caused by the short duration of experiment. Hence, the benefits of an ALT system on animal performance are expected after more study years due to a higher ANPP compared to CON plots. However, the experiment did not show any advantage of ROT on animal productivity due to the fact that herbage production was identical between years. Furthermore, extended grazed periods per year must cause a lower herbage allowance due to a continuously reduction in herbage mass on offer, which will, decrease feed intake and LWG of animals. Therefore, not only the duration of grazing years but also the duration of grazing within a year might alter the GS effects. A relative long grazing period per year might more clearly show the advantage of resting periods in sophisticated GS.

NRC (1975) stated that the requirement for metabolizable energy for maintenance and growth differs according to the age and species. Furthermore, Blaxter and Wainman (1964) and Garrett et al. (1959) concluded that the utilization of energy for maintenance and gain were different between sheep and cattle. Moreover, the capability for selecting of plants and plant parts is different between animals (Grant et al., 1985) and Animut et al. (2005) showed that diets of goats contain higher proportions of forbs than in sheep. Similarly, Kitessa and Nicol (2001) stated that the greater ability of sheep for selection might explain why LWG of sheep were similar, whereas for cattle higher at ROT than at CON in a cattle-sheep-co-grazed temperate pasture. Therefore, the choice of animal species and age could affect the effects of GS on animal feed intake and LWG, irrespective of herbage botanical composition and herbage quality.

As discussed above, many factors may alter the effects of GS on herbage production, nutritional quality of herbage, feed intake, and performance of grazing animals. Climate conditions in the studied ecosystem, the botanical composition of the grassland vegetation, the species and age of animal used, the SR, and the duration of grazing season should be considered when comparing GS effects. According to results of the present study and the literature review (Table 4.1), positive effects of sophisticated GS might be more pronounced under high SR, at a relative short duration of resting periods, and at a long-term grazing.

4.2 Choice of measured parameters

Schönbach et al. (2011) showed a higher ANPP of herbage at ALT grazing plots, however no improvements in dOM, feed intake as well as LWG of sheep were found after two years grazing in the present study (Chapter 3). Lower metabolizable energy concentration of herbage on offer was observed in 2008 while animal feed intakes were nearly identical (Chapter 2). Similarly, Davies et al. (2001) reported that herbage mass was higher at ALT than at CON while LWG of sheep was similar at both GS (Table 4.1). This might be explained by the ability of grazing animals to maintain their feed intake and thus, ingest enough energy for maintenance and growth by increasing grazing time (Animut et al., 2005; Lin et al., 2011) even at lower herbage allowances. Hence, animals may change their grazing behavior to compensate for the negative impacts of heavy grazing pressure. However, many studies compared effects of GS either on herbage (Clarke et al., 1943; Reardon and Merrill, 1976; Grant et al., 1987; Dowling et al., 1996) or on animal (Holechek et al., 1987; Heitschmidt et al., 1990) parameters while only in few studies both aspects were simultaneously considered (Table 4.1). Neither animal productivity nor ecological parameters can be ignored when a sustainable utilization of grasslands is intended.

4.3 Grazing systems for Inner Mongolia

As mentioned in Chapter 1, degradation and desertification of grassland in the Inner Mongolian steppe increased in last 5 decades due to the increasing human population and thus the decrease in the available grassland area for grazing (Lu et al., 2005; Zhao et al., 2005; Jiang et al., 2006). Hence, the purpose of the project "Matter fluxes of grasslands in Inner Mongolia as influenced by stocking rate" was to test the concepts discussed to affect a sustainable grassland utilization in Inner Mongolia. It could be shown that ANPP increased by 15% in response to an ALT grassland use when compared to CON grazing at moderate to heavy grazing intensity. Moreover, herbage ground cover was 35% higher at ALT than at CON under grazing intensity after 4 study years (Schönbach et al., 2011). Similarly, ALT did not change the species composition of herbage on offer while the above biomass of *Leymus chinensis*, one of the dominant species in the study area, declined with increasing SR in CON grazing plots (Wan et al., 2011). However, results presented in Chapter 3 did not show any advantage of ALT regarding animal feed intake and performance. As discussed in Section 4.2, herbage parameters cannot be ignored when considering long - term grasslands production. The positive effects of ALT on ANPP and herbage botanical composition were evidenced by previous data, which might lead to higher herbage allowances and herbage quality available for grazing. Therefore, although results in the present study did not show any positive effects of ALT on animal feed intake and LWG, it is still recommend that ALT systems should be applied in the Inner Mongolian steppe, since it may lead to higher LWG per area in the long-term due to preventing soil erosion compared with the current grassland use system. Herbage quality, dOM as well as animal feed intake and LWG were lower at ROT than at CON across the 4 years while herbage mass was similar at both GS (Chapter 2). Furthermore, it is discussed that grazing under higher SR might show the positive effects of ROT on animal performance. However, as mentioned in Chapter 1, increasing SR led to degradation and desertification of grassland in Inner Mongolia (Jiang et al., 2006). Hence, ROT cannot be seen as a sophisticated GS which maintain or even increase long term grassland production and animal productivity in this region. It has to be mentioned that in addition to the factors discussed above, grazing time, SR, animal species and age, and climate conditions should be considered as well when sophisticated and region-specific GS are developed. Furthermore, the change of the GS is solely not sufficient to resolve the ecological problems in Inner Mongolia. Combing the grassland protection policies mentioned in Chapter 1 with a suitable GS may save the grassland and maintain agricultural production in the future.

4.4 References

Allan, B. E., 1985. Grazing management on pasture and animal production from oversown tussock grassland. Proc. N. Z. Grassland Assoc.46, 119-125.

Allan, B. E., 1997. Grazing management of oversown tussock country 3. Effects on liveweight and wool growth of Merino wethers. N. Z. J. Agric. Res. 40, 437-447.

Alvarez-Rodriguez, J., Sanz, A., Delfa, R., Revilla, R., Joy, M., 2007. Performance and grazing behaviour of Churra Tensina sheep stocked under different management systems during lactation on Spanish mountain pastures. Livest. Sci.107, 152-161.

Animut, G., Goetsch, A. L., Aiken, G. E., Puchala, R., Detweiler, G., Krehbiel, C. R., Merkel, R. C., Sahlu, T., Dawson, L. J., Johnson, Z. B., Gipson, T. A., 2005. Grazing behavior and energy expenditure by sheep and goats co-grazing grass/forb pastures at three stocking rates. Small Ruminant Res.59, 191-201.

Bai, Y., Wu, J., Xing, Q., Pan, Q., Huang, J., Yang, D., Han, X., 2008. Primary production and rain use efficiency across a precipitation gradient on the Mongolia Plateau. Ecology 89, 2140-2153.

Bailey, D. W., 2004. Management strategies for optimal grazing distribution and use of arid rangelands. J. Anim. Sci.82 E-Suppl, E147-153.

Blaxter, K. L., Wainman, F. W., 1964. The utilization of the energy of different rations by sheep and cattle for maintenance and for fattening. J. Agr. Sci.63, 113-128.

Briske, D. D., Derner, J. D., Brown, J. R., Fuhlendorf, S. D., Teague, W. R., Havstad, K. M., Gillen, R. L., Ash, A. J., Willms, W. D., 2008. Rotational grazing on rangelands: Reconciliation of perception and experimental evidence. Rangeland Ecol. Manage.61, 3-17.

Campling R. C., Maclusky, D. S., Holmes, W., 1958. Studies in grazing management. VI. The influence of free - and strip - grazing and of nitrogenous fertilizers on production from dairy cows. J. Agr. Sci.51, 62-69.

Clarke, S. E., Tisdale, E. W., Skoglund, N. A., 1943. The effects of climate and grazing practices on short-grass prairie vegetation. C. Dept. Agr. Tech. Bull. 46, 53.

Davies, H. L., Southey, I. N., 2001. Effects of grazing management and stocking rate on pasture production, ewe liveweight, ewe fertility and lamb growth on subterranean clover-based pasture in Western Australia. Austr. J. Exp. Agr.41, 161-168.

Derner, J. D., Hart, R. H., Smith, M. A., Waggoner, J. W., 2008. Long-term cattle gain responses to stocking rate and grazing systems in northern mixed-grass prairie. Livest. Sci.117, 60-69.

Dowling, P. M., Kemp, D. R., Michalk, D. L., Klein, T. A., Millar, G. D., 1996. Perennial grass response to seasonal rests in naturalised pastures of central New South Wales. Rangeland J.18, 309-326.

Fynn, R. W. S., O'Connor, T. G., 2000. Effect of stocking rate and rainfall on rangeland dynamics and cattle performance in a semi-arid savanna, South Africa. J. Appl. Ecol.37, 491-507.

Garrett, W. N., Meyer, J. H., Lofgreen, G. P., 1959. The comparative energy requirements of sheep and cattle for maintenance and gain. J. Anim. Sci.18, 528-547.

Grant, S. A., Suckling, D. E., Smith, H. K., Torvell, L., Forbes, T. D. A., Hodgson, J., 1985. Comparative studies of diet selection by sheep and cattle-the hill grasslands. J. Ecol.73, 987-1004.

Hacker, R. B., Richmond, G. S., 1994. Simulated Evaluation of Grazing Management-Systems for Arid Chenopod Rangelands in Western-Australia. Agric. Sys.44, 397-418.

Hafley, J. L., 1996. Comparison of Marshall and Surrey ryegrass for continuous and rotational grazing. J. Anim. Sci.74, 2269-2275.

Heitschmidt, R. K., Dowhower, S. L., Walker, J. W., 1987. Some effects of a rotational grazing treatment on quantity and quality of available forage and amount of ground litter. J. Range Manage.40, 318-321.

Heitschmidt, R. K., Conner, J. R., Canon, S. K., Pinchak, W. E., Walker, J. W., Dowhower, S. L., 1990. Cow/calf production and economic returns from yearlong continuous, deferred rotation and rotational grazing treatments. J. Prod. Agr.3, 92-99.

Hodgson, J., 1985. The control of herbage intake in the grazing ruminant. Proc. Nutr. Soc.44, 339-346.

Holechek, J. L., Berry, T. J., Vavra, M., 1987. Grazing system influences on cattle performance on mountain-range. J. Range Manage.40, 55-59.

Holmes, W., Osman, H. E. S., 1960. The feed intake of grazing cattle. I. Feed intake of dairy cows on strip and free grazing. Anim. Prod.2, 131-139.

Jamieson, W. S., Hodgson, J., 1979. The effect of daily herbage allowance and sward characteristics upon the ingestivebehaviour and herbage intake of calves under strip-grazing management. Grass Forage Sci.34, 261-271.

Jiang, G. M., Han, X. G., Wu, J. G, 2006. Restoration and management of the Inner Mongolia grassland require a sustainable strategy. Ambio 35, 269-270.

Joy, M., Alvarez-Rodriguez, J., Revilla, R., Delfa, R., Ripoll, G., 2008. Ewe metabolic performance and lamb carcass traits in pasture and concentrate-based production systems in Churra Tensina breed. Small Ruminant Res. 75, 24-35.

Kitessa, S. M., Nicol, A. M., 2001. The effect of continuous or rotational stocking on the intake and live-weight gain of cattle co-grazing with sheep on temperate pastures. Anim. Sci.72, 199-208.

Lin, L. J., Dickhoefer, U., Müller, K., Wurina, Susenbeth, A., 2011. Grazing behavior of sheep at different stocking rates in the Inner Mongolian steppe, China. Appl. Anim. Behav. Sci.129, 36-42.

Lodge, G. M., Murphy, S. R., Harden, S., 2003. Effects of grazing and management on herbage mass, persistence, animal production and soil watercontent of native pastures. 1. A redgrass-wallaby grass pasture, Barraba, North West Slopes, New South Wales. Austr. J. Exp. Agr.43, 875-890.

Lu, Z. J., X. S. Lu, and X. P. Xin. 2005. Present situation and trend of grassland desertification of North China. Acta Agrestia Sin.13, 24-27 (in Chinese).

Manley, W. A., Hart, R. H., Samuel, M. J., Smith, M. A., Waggoner, J.

W., Manley, J. T., 1997. Vegetation, cattle, and economic responses to grazing strategies and pressures. J. Range Manage.50, 638-646.

Miller, D. R., Dean, G. J., Ball, P. D., 2010. Influence of end - grazing forage residual and grazing management on lamb growth performance and crop yield from irrigated dual-purpose winter wheat. Anim. Prod. Sci.50, 508-512.

Nie, Z. N., Zollinger, R. P., 2011. Impact of deferred grazing and fertilizer on plant population density, ground cover and soil moisture of native pastures in steep hill country of southern Australia. Grass Forage Sci.67, 231-242.

NRC, 1975. Nutrient requirements of sheep, fifth ed. National Academy Press, Washington, DC.

Patton, B. D., Dong, X. J., Nyren, P. E., Nyren, A., 2007. Effects of grazing intensity, precipitation, and temperature on forage production. Rangeland Ecol. Manag.60, 656-665.

Popp, J. D., McCaughey, W. P., Cohen, R. D. H., 1997. Effect of grazing system, stocking rate and season of use on diet quality and herbage availability of alfalfa-grass pastures. Can. J. Anim. Sci.77, 111-118.

Reardon, P.O., Merrill, L. B., 1976. Vegetative response under various grazing management systems in the Edwards Plateau of Texas. J. Range Manage. 29, 195-198.

Schönbach, P., Wan, H., Schiborra, A., Gierus, M., Bai, Y., Müller, K., Glindemann, T., Wang, C., Susenbeth, A., Taube, F., 2009. Short-term management and stocking rate effects of grazing sheep on herbage quality and productivity of Inner Mongolia steppe. Crop Pasture Sci. 60, 963-974.

Schönbach, P., Wan, H. W., Gierus, M., Bai, Y. F., Müller, K., Lin, L. J., Susenbeth, A., Taube, F., 2011. Grassland responses to grazing: effects of grazing intensity and management system in an Inner Mongolian steppe ecosystem. Plant Soil 340, 103-115.

Simon, J. R. W., Graeme, C. W., David, G. M., 1995. Optimal grazing of a multi-paddock system using a discrete time model. Agric. Sys.48, 119-139.

Virgona, J. M., Avery, A. L., Graham, J. F., Orchard, B. A., 2000. Effects of grazing management on phalaris herbage mass and persistence in summer-dry environments. Aust. J. Exp. Agric.40, 171-184.

Wan, H. W., Bai, Y. F., Schönbach, P., Gierus, M., Taube, F., 2011. Effects of grazing management system on plant community structure and functioning in a semiarid steppe: scaling from species to community. Plant Soil 340, 215-226.

Wang, C. J., Tas, B. M., Glindemann, T., Müller, K., Schiborra A., Schönbach P., Gierus, M., Taube, F., Susenbeth, A., 2009. Rotational and continuous grazing of sheep in the Inner Mongolian steppe of China. J. Anim. Physiol. Anim. Nutr.93, 245-252.

Warner, J. R., Sharrow, S. H., 1984. Set stocking, rotational grazing and forward rotational grazing by sheep on western oregon hill pastures. Grass Forage Sci.39, 331-338.

Zhao, H. L., Zhao, X. Y., Zhou, R. L., Zhang, T. H., Drake, S., 2005. Desertification processes due to heavy grazing in sandy rangeland, Inner Mongolia. J. Arid Environ.62, 309-319.

Chapter 5 General conclusions

5 General conclusions

Herbage quality, digestibility of ingested organic matter, feed intake, and live weight gain of sheep decreased at rotational (ROT) grazing system. This might be due to greater-stocking densities at ROT paddocks, which might have limited the selected behavior of grazing sheep and thus reduced the nutritional quality of the diet. Results showed that ROT is not a suitable management system for the Inner Mongolian steppe to maintain or even increase herbage production and animal performance. Ground cover and above-ground net primary production (ANPP) of herbage were higher at alternating (ALT) than continuous (CON) grazing at a moderate to heavy grazing intensity. In contrast thereto, grazing effects on animal parameters were similar at both systems. The limited duration of the present studies, the variable distribution and amount of rainfall, and there lative short term duration might explain why the ALT neither improved feed intake and live weight gain of sheep nor compensated for the negative effects of moderate to high stocking rate on animal performance in 2009. It is recommended that ALT should be applied in the Inner Mongolian steppe due to the advantage of ALT on herbage ground cover, ANPP as well as botanical composition. By this, ALT may increase the long-term animal productivity. In view of previous literature, it is concluded that long term duration of grazing practice and sufficient precipitation are benefits to show the advantages of sophisticated grazing systems in Inner Mongolia. Moreover, a study should measure herbage and animal parameters simultaneously and be performed for a long duration to comprehensively compare the effects of different grazing systems.

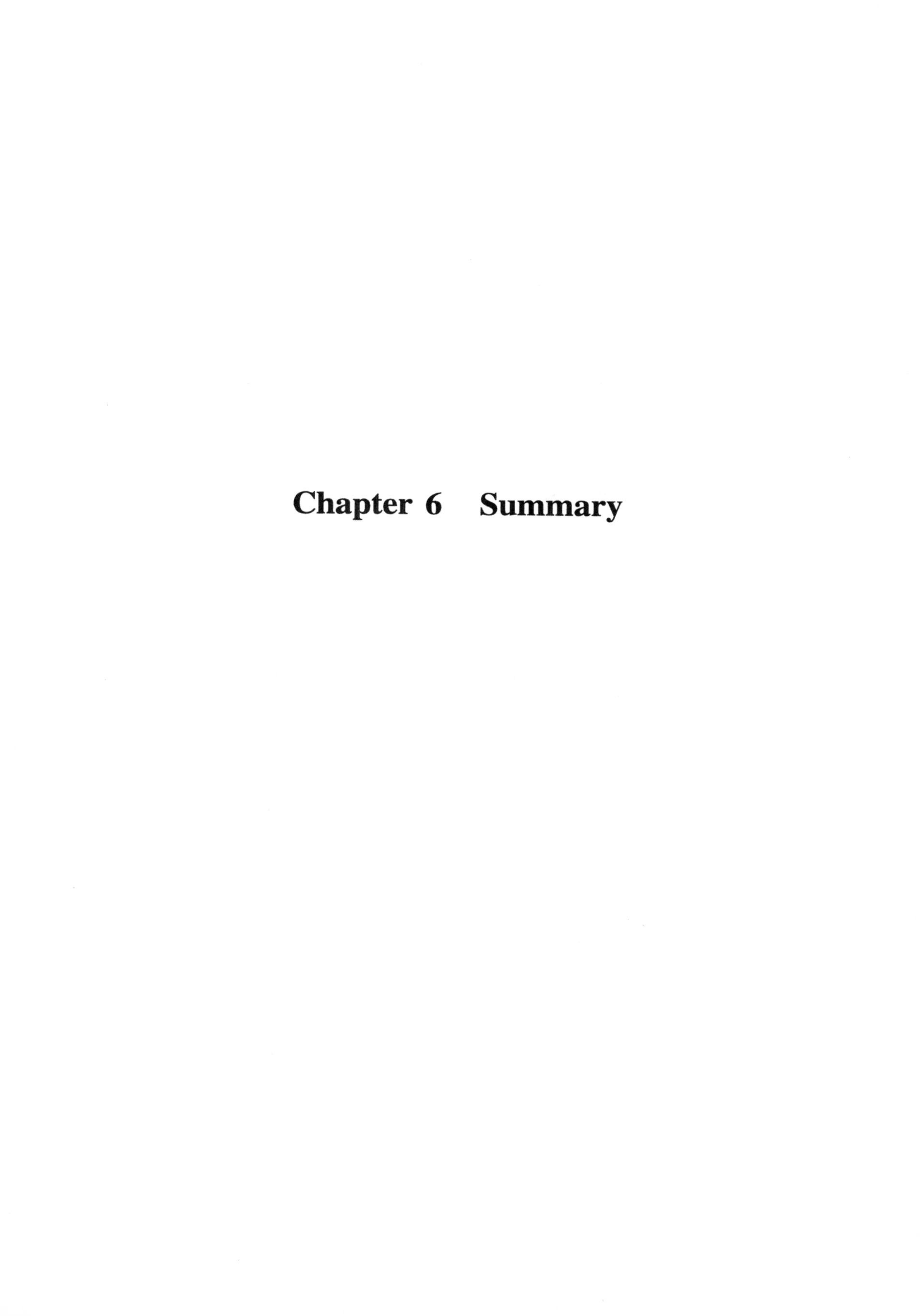

Chapter 6 Summary

6 Summary

Increasing human population decreased the available grassland area per sheep in the Inner Mongolian steppe in last 60 years. Due to the sedentarization of nomadic families, grassland close to settlements is currently used for intensive sheep and cattle grazing, while distant areas are only used for hay-making in this region. The consequences of this grassland utilization change are the degradation and desertification of the Inner Mongolian steppe. World-wide, grassland managers considered that sophisticated grazing management systems (GS) are an important tool for maintaining or even increasing long-term grassland production and animal performance. Therefore, many studies compared herbage production, herbage quality, and change of species composition at sophisticated GS with continuous (CON) grazing. However, few studies measured the influence of GS on digestibility of ingested herbage organic matter (dOM) and feed intake or only evaluated live weight gain (LWG) of animals. Hence, the main objectives of the thesis were to analyze effects of GS on dOM, organic matter intake (OMI), and LWG of grazing sheep in the Inner Mongolian steppe.

Two sophisticated GS's, a rotational (ROT) and an alternating (ALT) GS, were tested with regard to their effects on dOM, feed intake, and LWG of sheep. In ROT, sheep sequentially grazed in four 0.5-ha paddocks for 10 d each at a moderate stocking rate of 4.2 sheep/ha. Instead, in CON sheep grazed the whole plots throughout the entire grazing season. Across the four study years from 2005 to 2008, dOM, OMI, and LWG were lower at ROT than at CON. However, differences could not be observed every year. Hence, it can be concluded that ROT was not superior or even inferior to CON in Inner Mongolia. In the ALT system, grazing and hay-making were alternated annually between two adjacent plots, while sheep grazed the same plots every year in CON. Six different grazing intensities (GI) were established from very light to very heavy grazing. Data collected in 2009 and 2010 showed that GS did not affect dOM,

OMI, and LWG of sheep. Nevertheless, in view of published data of the same project, ALT might improve the long term productivity and feeding value of the steppe vegetation. Hence, it is recommended that ALT might enhance revenues and ecological sustainability of grassland when compared to the common practice of CON grazing at very high stocking rates.

Appendix 1　AOAC Official Method 942.05 Ash of Animal Feed First Action 1942 Final Action

Weigh 2 g test portion into porcelain crucible and place in a temperature controlled furnace preheated to 600℃. Hold at this temperature 2 h. Transfer crucible directly to desiccator, cool, and weigh immediately, reporting percent ash to first decimal place.

$$\% \text{ (w/w) ash} = \frac{\text{weight of test portion, g} - \text{weight loss on ashing, g}}{\text{weight of test portion, g}} \times 100$$

References:

JAOAC 25, 857 (1942); 26, 220 (1943).

Appendix 2 AOAC Official Method 2001.11 Protein (Crude) in Animal Feed, Forage (Plant Tissue), Grain, and Oilseeds Block Digestion Method Using Copper Catalyst and Steam Distillation into Boric Acid First Action 2001

[Applicable to the determination of 0.5% ~ 50% Kjeldahl N (3% ~ 300% equivalent crude protein) in forage, animal feed and pet food, grain, and oilseeds, and applicable to the same matrixes as 976.05 (4.2.05), 976.06 (4.2.06), 984.13 (4.2.09), 988.05 (4.2.03), and 990.02 (4.2.07); the method does not measure oxidized forms of N or heterocyclic N compounds.]

See Tables 2001.11A and B for the results of the interlaboratory study, expressed on a protein basis (N×6.25), supporting acceptance of the method.

A. Principle

The material is digested in H_2SO_4 to convert the protein N to $(NH_4)_2SO_4$ at a boiling point elevated by the addition of K_2SO_4 with a Cu catalyst to enhance the reaction rate. Ammonia is liberated by alkaline steam distillation and quantified titrimetrically with standardized acid. Aluminum block heaters increase the efficiency of the digestion.

The digest must contain residual H_2SO_4 to retain the NH_3. Water is added manually or automatically to the digest to avoid mixing concentrated alkali with concentrated acid and to prevent the digest from solidifying. Concentrated NaOH is added to neutralize the acid and make the digest basic, and the liberated NH_3 is distilled into a boric acid solution and titrated with a stronger standardized acid, HCl, to a colorimetric endpoint. The same endpoint detection system (e. g. , indicator, wavelength) must be used for the standardization of the HCl and for the analyte.

The analyte is referred to as "crude" protein because the method determines N, a component of all proteins. In addition, N from sources other than true protein is also determined. (Additional digestion procedures must be used in order to include N from nitrate.) The amount of protein in most materials is calculated by multiplying % N by 6. 25, because most proteins contain 16% N.

The H_2SO_4 and NaOH used are in concentrated form and are highly corrosive. Wear gloves and eye protection while handling the chemicals. Do not mix concentrated acid and NaOH directly. If chemicals are splashed on the skin or in the eyes, flush with copious amounts of water. Seek medical attention. Do not breathe the sulfur oxide fumes produced during digestion.

B. Apparatus

(a) *Digestion block.* —Aluminum alloy block with adjustable temperature device for measuring and controlling block temperature (Tecator Digestion System 20, 1015 Digestor, Foss North America, 7682 Executive Dr, Eden Prairie, MN 55344, USA; +1-952-974-9892, Fax: +1-952-974-9823, info@ fossnorthamerica. com; or equivalent).

(b) *Digestion tubes.* —250 mL.

(c) *Distillation units.* — (1) *For steam distillation.* —Foss Tecator 2200, or equivalent, to accept 250 mL digestion tubes and 500 mL titration flasks. (2) *For steam distillation and autotitration.* —Foss Tecator 2300, or equivalent.

(d) *Titration flask.* —500 mL graduated Erlenmeyer flask (for collection and titration of distillate).

(e) *Fume exhaust manifold.* —With Teflon ring seals, connected to a water aspirator in a hooded sink.

(f) *Weighing paper.* —Low N, Alfie Packers No. 201 (Alfie Packers, Inc., 8901 J St, Ste 10, Omaha, NE 68127, USA), or Fisher 09-898-12A, 3×3 in. (76 mm × 76 mm), or equivalent.

(g) *Pipetting dispenser.* —25 mL, adjustable volume, attached to a 5 pint (2.4 L) acid bottle.

C. Reagents

(a) *Sulfuric acid.* —Concentrated, 95%~98% H_2SO_4, reagent grade.

(b) *Catalyst.* —7.0 g K_2SO_4 + 0.8 g $CuSO_4$. (Commercially available in tablet form as 3.5 g K_2SO_4 and 0.4 g $CuSO_4$ per tablet.)

(c) *Sodium hydroxide solution.* —40% (w/w) NaOH, low N ($\leqslant$5 μg N/g).

(d) *Methyl red indicator solution.* —Dissolve 100 mg methyl red in 100 mL methanol.

(e) *Bromocresol green indicator solution.* —Dissolve 100 mg bromocresol green in 100 mL methanol.

(f) *Boric acid solution.* —4% (w/v). Dissolve 400 g H_3BO_3 in 5~6 L hot deionized water. Mix and add more hot deionized water to a volume of about 9 L. Cool to room temperature, add 100 mL bromocresol green solution and 70 mL methyl red solution, and dilute to a final volume of 10 L. Adjust to obtain a positive blank of 0.05~0.15 mL with 30 mL H_3BO_3 solution, using 0.1M NaOH (to increase blank) or 0.1M HCl (to decrease blank). Commercially available.

(g) *Boric acid solution.* —1% (w/v). (Optional trapping solution for titrators that automatically begin titration when distillation begins.) Dissolve 100 g H_3BO_3 in 5~6 L hot deionized water, mix, and add more hot deionized water to a volume of about 9 L. Cool to room temperature, add 100 mL bromocresol green solution and 70 mL methyl red solution, and dilute to a final volume of 10 L. Commercially available.

(h) *Hydrochloric acid standard solution.* —0.1000M. Prepare as in 936.15 (*see* A. 1.06) or use premade solution of certified specification range 0.0995 ~ 0.1005M, and use 0.1000M for calculation. Commercially available.

(i) *Reference standards.* —Ammonium sulfate, tryptophan, lysine · HCl, or glycine *p*-toluenesulfonic acid, for use as standard; 99.9%.

(j) *Sucrose.* —N-free.

D. Preparation of Analytical Sample

Grind dry laboratory sample to fineness of grind (ca 0.7-1 mm), which gives a relative standard deviation (RSD) of ≤2.0% for 10 successive determinations of N in ground mixture of corn grain and soybeans (2 + 1). Fineness required to achieve this precision must be used for all dry mixed feeds and other nonuniform materials. Mix liquids to uniformity.

E. Determination

(a) *Digestion.* —Turn on block digestor and heat to 420℃. Weigh materials, as indicated below, recording each test portion weight (W) to the nearest mg for weights of ≥1 g, and to the nearest 0.1 mg for weights of <1.0 g. Do not exceed 1.2 g. For materials with 3% ~ 25% protein, weigh approximately 1.0 g test portion; with 25% ~ 50% protein, approximately 0.5 g test portion; and >50% protein, approximately 0.3 g test portion.

(1) *Dry feed, forage, cereal, grain, oilseeds.* —Weigh 1 g test portion of ground, well-mixed test portion onto a tared, low N weighing paper. Fold paper around material and drop into a numbered Kjeldahl tube.

(2) *Liquid feed.* —Weigh slightly >1 g test portion of well-mixed analytical

sample into a small tared beaker. Quantitatively transfer to a numbered Kjeldahl tube with <20 mL deionized water. Alternatively, weigh slightly >1 g well-mixed test portion into a small tared beaker. Transfer to a numbered Kjeldahl tube and reweigh beaker. The differential weight loss corresponds to the amount of test portion actually transferred to the tube.

(b) *Standards.* —Perform quality control analysis and analyses of standards with each batch. The standards available from Hach Co. (PO Box 389, Loveland, CO 80539, USA; +1-800-227-4224 or +1- 970-669-3050), Sigma (St. Louis, MO), J. T. Baker (Phillipsburg, NJ), the National Institute of Standards and Technology (NIST; Gaithersburg, MD) are listed in Table 2001.11C.

The various ammonium salts and glycine*p*-toluenesulfonate serve primarily as a check on distillation efficiency and accuracy in titration steps because they are digested very readily. Lysine and nicotinic acid *p*-toluenesulfonate serve as a check on digestion efficiency because they are difficult to digest.

Include a reagent blank tube containing a folded low N weighing paper with each batch.

(c) *Digestion.* —Add 2 catalyst tablets to each tube. Add 12 mL H_2SO_4 to each tube, using pipetting dispenser; add 15 mL for high fat materials (>10% fat). Mixtures may be held overnight at this point. If mixture foams, slowly add 3 mL 30%～35% H_2O_2. Let reaction subside in perchloric acid fume hood or in exhaust system.

Attach heat side shields to tube rack. Place fume manifold tightly on tubes, and turn water aspirator on completely. Place rack of tubes in preheated block. After 10 min, turn water aspirator down until acid fumes are just contained within exhaust hood. A condensation zone should be maintained within the tubes. After bulk of sulfur oxide fumes are produced during initial stages of digestion, reduce vacuum source to prevent loss of H_2SO_4. Digest additional 50 min. Total digestion time is approximately 60 min.

Turn digestor off. Remove rack of tubes with exhaust still in place, and put in the stand to cool for 10～20 min. Cooling can be increased by using commercial air blower or by placing in hood with hood sash pulled down to increase airflow across tubes. When fuming has stopped, remove manifold, and shut off aspirator. Remove aside shields. Let tubes cool. Wearing gloves and eye protection, predilute digests manually before distilling. Carefully add a few milliliters of deionized water to each tube. If spattering occurs, the tubes are too hot. Let them cool for a few more minutes. Add water to each tube to a total volume of approximately 80 mL (liquid level should be about half way between the 2 shelves of the tube rack). This is a convenient stopping point.

If digest solidifies, place tube containing diluted digest in block digester, and carefully warm with occasional swirling until salts dissolve. If distilling unit equipped with steam addition for equilibration is used, the manual dilution steps can be omitted. About 70 mL deionized water is then automatically added during the distillation cycle.

(d) *Distillation.* —Place 40% NaOH in alkali tank of distillation unit. Adjust volume dispensed to 50 mL. Attach digestion tube containing diluted digest to distillation unit, or use automatic dilution feature, if available. Place graduated 500 mL Erlenmeyer titration flask containing 30 mL H_3BO_3 solution with indicator on receiving platform, and immerse tube from condenser below surface of H_3BO_3 solution. (When an automatic titration system is used that begins titration immediately after distillation starts, 1% H_3BO_3 may be substituted.) Steam distill until ≥150 mL distillate is collected (≥180 mL total volume). Remove receiving flask. Titrate H_3BO_3 receiving solution with standard 0.1000M HCl to violet endpoint (just before the solution goes back to pink). Lighted stir plate may aid visualization of endpoint. Record milliliters of HCl to at least the nearest 0.05 mL.

This is done automatically by using a steam distiller with automatic titration. Follow the manufacturer's instructions for operation of the specific distiller or distiller/titrator.

F. Verification of Nitrogen Recovery

Run N recoveries to check accuracy of procedure and equipment.

(a) *Nitrogen loss.* —Use 0.12 g $(NH_4)_2SO_4$ and 0.67 g sucrose per flask. Add all other reagents as in E, and distill under same conditions as in E. Recoveries must be ≥99%.

(b) *Distillation and titration efficiency.* —Distill 0.12 g $(NH_4)_2SO_4$, omitting digestion. Recoveries must be ≥99.5%.

(c) *Digestion efficiency.* —Use 0.3 g acetanilide or 0.18 g tryptophan, with 0.67 g sucrose per flask. Add all other reagents as stated in E. Digest and distill under same conditions as used for a determination. Recoveries must be ≥98%.

G. Calculations

$$\text{Kjeldahlnitrogen}, \% = \frac{(V_S - V_b) \times M \times 14.01}{W \times 10}$$

$$\text{Crude protein}, \% = \%\text{Kjeldahl N} \times F$$

Where V_S = volume (mL) of standardized acid used to titrate a test; V_B = volume (mL) of standardized acid used to titrate reagent blank; M = molarity of standard HCl; 14.01 = atomic weight of N; W = weight (g) of test portion or standard; 10 = factor to convert mg/g to percent; and F = factor to convert N to protein.

F factors are 5.70 for wheat, 6.38 for dairy products, and 6.25 for other feed materials.

Reference:

J. AOAC Int. (future issue).

Table 2001.11A: Interlaboratory study results for the determination of crude protein by block digestion with a copper catalyst and distillation into 4% boric acid

Table 2001.11B: Interlaboratory study results for the recovery of nitrogen from standard compounds by block digestion with a copper catalyst and distillation into boric acid

Table 2001.11C: Standards

Table 2001.11A: Interlaboratory study results for the determination of crude protein by block digestion with a copper catalyst and distillation into 4% boric acid

ID	No. oflabs[a]	Mean, %	RSD_r, %	RSD_R, %	HORRAT
Protein block	10 (1)	40.19	0.45	0.76	0.333
Swine pellets	10 (1)	37.04	0.47	0.60	0.256
Corn silage	11	7.10	1.64	2.16	0.726
Grass hay	11	7.11	1.94	1.94	0.650
Fish meal	11	64.67	0.73	0.98	0.460
Dog food	11	24.50	0.87	0.91	0.369
Chinchilla food	11	18.01	0.89	0.99	0.383
Albumin	10 (1)	79.14	0.4	0.44	0.212
Birdseed	11	13.48	0.88	1.29	0.475
Meat and bone meal	11	50.06	1.90	1.90	0.857
Milk replacer	11	20.78	1.39	1.39	0.550
Soybeans	9 (2)	38.76	0.49	0.54	0.236
Sunflower seeds	11	17.43	2.38	2.38	0.916
Legume hay	11	18.81	1.45	1.45	0.565

[a] Each value is the number of laboratories retained after elimination of outliers; each value in parentheses is the number of laboratories removed as outliers.

Table 2001.11B: Interlaboratory study results for the recovery of nitrogen from standard compounds by block digestion with a copper catalyst and distillation into boric acid

Compound	No. oflabs[a]	Theoretical yield, % N	Avg. found, % N	Avg. rec., %	RSD_R, %	HORRAT
Acetanilid	10 (0)	10.36	10.37	100.1	1.50	0.53
Lysine · HCl	10 (0)	15.34	13.32	86.8	4.16	1.53
Tryptophan	10 (0)	13.72	13.55	98.8	1.04	0.39

[a] Each value is the number of laboratories retained after elimination of outliers; each value in parentheses is the number of laboratories removed as outliers.

Table 2001.11C: Standards

Standard	Approximate weight, g	Theoretical yield, % N
Ammonium p-toluenesulfonate (Hach 22779-24)	0.5	7.402
Glycine*p*-toluenesulfonate (Hach 22780-24)	0.6	5.665
Glycine p-toluenesulfonate (Hach 22780-24)	0.6	5.665
Nicotinic acid p-toluenesulfonate (Hach 22781-24)	0.2	4.743
Lysine monohydrochloride (Sigma L-5626)	0.1	15.34
Acetanilide (Baker A068-05)	0.3	10.36
Tryptophan (Sigma T 8659)	0.2	13.72
Ammonium salts Diammonium hydrogen phosphate (100% assay)	0.2	21.21
Ammonium chloride (100% assay)	0.2	26.18
Ammonium sulfate (100% assay)	0.2	21.2
Ammonium dihydrogen phosphate (NIST 200)	0.3	12.18
Citrus leaves (NIST 1572)	1	2.86
Urea (NIST 2141)	0.1	46.63

Appendix 3 AOAC Official Method 920.39 Fat (Crude) or Ether Extract in Animal Feed First Action 1920 Final Action

Use method A or C for feed ingredients and mixed feeds other than ①baked and/or expanded, ②dried milk products, ③containing urea, or ④mixed feeds that have at least 20% of crude fat derived from baked and/or expanded, or dried milk products.

A. Indirect Method

Determine moisture as in 934.01 ; then extract dried substance as in C, and dry again. Report loss in weight as ether extract.

Direct Method

B. Reagent

Anhydrousether. —Wash commercial ether with 2 or 3 portions H_2O, add solid NaOH or KOH, and let stand until most of H_2O is abstracted from the ether. Decant into dry bottle, add small pieces of carefully cleaned metallic Na, and let stand until H_2 evolution ceases. Keep ether, thus dehydrated, over metallic Na in loosely stoppered bottles.

C. Determination

(Large amounts H_2O – soluble components such as carbohydrates, urea, lactic acid, glycerol, and others may interfere with extraction of fat; if present, extract 2 g test portion on small paper in funnel with five 20 mL portions H_2O prior to drying for ether extraction.)

Extract ca 2 g test portion, dried as in 934.01 , with anhydrous ether. Use thimble with porosity permitting rapid passage of ether. Extraction period may vary from 4 h at condensation rate of 5 ~ 6 drops/s to 16 h at 2 ~ 3 drops/s. Dry extract 30 min at 100℃, cool, and weigh.

References:

JAOAC 64, 351 (1981); 65, 289 (1982).

ISO 6492: 1999 (E).

Appendix 4 AOAC Official Method 934.01 Loss on Drying (Moisture) at 95~100℃ for Feeds Dry Matter on Oven Drying at 95~100℃ for Feeds First Action 1934 Final Action Codex-Adopted-AOAC Method*

(This method is not applicable to feeds containing >5% urea.)

Dry test portion containing ca 2 g dry material to constant weight at 95~100℃ under pressure≤100 mm Hg (ca 5 h). For feeds with high molasses content, use temperature ≤70℃ and pressure ≤50 mm Hg. Use covered Al dish ≥50 mm diameter and 40 mm deep. Report loss on drying (LOD) as estimate of moisture content.

Calculations

$$\%\ (w/w)\ \text{LOD} = \%\ (w/w)\ \text{moisture} = 100 \times \frac{wtlossondrying,\ g}{wttestportion,\ g}$$

$$\%\ \text{Dry matter} = 100 - \%\ \text{LOD}$$

References:

JAOAC 17, 68 (1934); 51, 467 (1968);

60, 322 (1977).

Revised: *March* 1998

* Adopted as Codex Defined Method (Type I) for gravimetry in the loss of drying in special foods.

Appendix 5 AOAC Official Method 978.10 Fiber (Crude) in Animal Feed and Pet Food Fritted Glass Crucible Method First Action 1978 Final Action 1979

A. Principle

Principle is same as in 962.09A, except sample solution is exposed to minimum vacuum needed to regulate filtration, and heating of sample solutions prevents gelling or precipitation of possible saturated solutions.

B. Apparatus and Reagents

See Reagents 962.09B (a), (b), and (f); *Apparatus* 962.09C (a), (c), (d), and (f), and in addition:

(a) *Filtration apparatus.* —System to permit application of minimum vacuum necessary for filtration and washing of each sample within 3 ~ 5 min. Each unit consists of reservoir manifold connected to ① H_2O aspirator through 120℃ stopcock, ②atmosphere through second stopcock with metering device, and ③ receptacle containing cone-shaped hard rubber gasket which provides vacuum seal with crucible. Vacuum gage attached to manifold indicates vacuum applied to crucible. Crucible can be heated before and during filtration by flow of hot H_2O in surrounding jacket. [For photograph of apparatus, *see JAOAC* 56, 1353 (1973). Filtration unit is available as Model 150 (now Model AS-2000) from Analytical Bio-Chemistry Laboratories, Inc. , 7200 ABC Ln,

Columbia, MO 65205.]

(b) *Crucible.* —Fritted glass, 50 mL, coarse porosity. Clean as follows: Brush, and flow hot tap H_2O into crucible to remove as much ash as possible. Submerge crucible in base solution, (c) (2), ≥5 min, remove, and rinse with hot tap H_2O. Submerge in HCl (1 + 1), (c) (1), ≥5 min, remove, and rinse thoroughly with hot tap H_2O followed by distilled H_2O. After 3~4 uses, back wash by inverting crucible on hard rubber gasket in filtration apparatus, and flowing near-boiling H_2O through crucible under partial vacuum.

(c) *Cleaning solutions.* — (1) *Acid solution.* —HCl (1 + 1). (2) *Base solution.* —Dissolve 5 g Na_2H_2EDTA, 50 g Na_2HPO_4 (technical grade), and 200 g KOH in H_2O, and dilute to 1 L. Storage in separate wide-mouth containers holding 2~3 L solution into which crucibles can be placed is convenient.

(d) *Filtering device.* —Modified California plastic Buchner (*see* Figure 962.09A). *See* 962.09C (d).

C. Determination

Extract 2 g ground test portion with ether or petroleum ether (initial boiling temperature, 35~38℃; dry-flask end point, 52~60℃; ≥95% distilling <54℃, and ≤60% distilling <40℃; specific gravity at 60°F, 0.630~0.660; evaporation residue ≤0.002% by weight). If fat is ≤1%, extraction may be omitted. Transfer to 600 mL reflux beaker, avoiding fiber contamination from paper or brush. Add 0.25~0.5 g bumping granules, followed by 200 mL near-boiling 1.25% H_2SO_4, 962.09B (a), solution in small stream directly on sample to aid in complete wetting of sample. Include beakers with no test portion (blank). Run 2 blanks for every 24 test samples. Place beakers on digestion apparatus at 5 min intervals and boil exactly 30 min, rotating beakers periodically to keep solids from adhering to sides. Near end of refluxing place California Buchner, 962.09C (d), previously fitted with No. 9 rubber stopper to provide vacuum seal, into filtration apparatus, and adjust vacuum to ca 25 mm Hg (735 mm pressure). At the end of refluxing, flow near-boiling H_2O through funnel to warm it; then decant liquid through funnel, washing solids into funnel with minimum of near-boiling H_2O. Filter to dryness, using 25 mm vacuum, and wash residue with four 40~50 mL portions near-boiling H_2O, filtering after each washing. Do not add wash to funnel under vacuum; lift funnel from apparatus when adding wash.

Wash residue from funnel into reflux beaker with near-boiling 1.25% NaOH, 962.09B (b), solution. Place beakers on reflux apparatus at 5 min intervals and reflux 30 min. Near end of refluxing, turn on filtration apparatus, place crucible, B (b), and adjust vacuum to ca 25 mm. Flow near-boiling H_2O through crucible to warm it. (Keep near-boiling H_2O flowing through jacket during filtration and washing.) At end of refluxing, decant liquid through crucible and wash solids into crucible with minimum of near-boiling H_2O. Increase vacuum as needed to maintain filtration rate. Wash residue once with 25~30 mL near-boiling 1.25% H_2SO_4 solution, and then with two 25~30 mL portions near-boiling H_2O, filtering after each washing. (Filtering and washing takes ca 3~5 min/test portion.) Do not add wash to crucible under vacuum.

Dry crucible with residue 2 h at 130℃ ± 2℃ or overnight at 110℃, cool in desiccator, and weigh (W_2). Ash 2 h at 550℃ ± 10℃, cool in desiccator, and weigh (W_3). Do not remove crucibles from furnace until temperature is ≤250℃, as fritted disk may be damaged if cooled too rapidly.

If needed for converting moisture basis, determine moisture of the test sample using a separate portion.

Crude fiber, % = loss in weight on ignition × 100/weight test portion, g

$$\text{Crude fiber, \%} = [(W_2 - W_3) - (B_2 - B_3)] / W_1 \times 100$$

Crude fiber on desired moisture basis (or desired dry matter basis),

$$\%\ (w/w) = C \times \frac{100 - \%\ \text{moisturebasisdesired}}{100 - \%\ \text{moistureintestsample}}$$

$$\text{or } C \times \frac{\%\ \text{dry matterbasisdesired}}{\%\ \text{dry matterintestsample}}$$

Where B_2 and B_3 are average weights of all blanks after oven drying and ashing, resepectively.

References:

JAOAC 61, 154 (1978).

Revised: *March* 1996

962.09A:

Crude fiber is loss on ignition of dried residue remaining after digestion of sample with 1.25% (w/v) H_2SO_4 and 1.25% (w/v) NaOH solutions under specific conditions. Method is applicable to materials from which the fat can be and is extracted to obtain a workable residue, including grains, meals, flours, feeds, fibrous materials, and pet foods.

962.09B:

(a) *Sulfuric acid solution.* —0.128M± 0.003M. 1.25 g H_2SO_4/100 mL. Concentration must be checked by titration.

(b) *Sodium hydroxide solution.* —1.25 g NaOH/100 mL, 0.313M ± 0.005M, free, or nearly so, from Na_2CO_3. Concentration must be checked by titration.

(f) *Bumping chips or granules.* —Broken Alundum crucibles or equivalent granules (RR Alundum 90 mesh, Norton Co. , 1 New Bond St, Worcester, MA 01606, USA) are satisfactory.

962.09C:

(a) *Digestion apparatus.* —With condenser to fit 600 mL beaker, and hot plate adjustable to temperature that will bring 200 mL H_2O at 25℃ to rolling boil in 15 min± 2 min. (Available from Labconco Corp. , 8811 Prospect Ave, Kansas City, MO 64132, USA.)

(c) *Desiccator.* —With efficient desiccant such as 4-8 mesh Drierite ($CaCl_2$ is not satisfactory).

(d) *Filtering device.* —With No. 200 Type 304 or 316 stainless steel screen (W. S. Tyler, Inc. , 8570 Tyler Blvd, Mentor, OH 44060, USA), easily washed free of digested residue. Either Oklahoma State filter screen (*see* Figure 962.09A; available from Labconco Corp.) or modified California plastic Buchner (*see* Figure 962.09B; consists of 2 piece polypropylene plastic funnel manufactured by Nalge Co. , 75 Panorama Creek Dr, PO Box 20365, Rochester, NY 14602, USA, Cat. No. 4280-0700, 70 mm [without No. 200 screen], or equivalent [also available from Labconco Corp.]. Seal screen to filtering surface of funnel, using small-tip soldering iron).

(f) *Liquid preheater.* —For preheating H_2O, 1.25% H_2SO_4, B (a), and 1.25% NaOH, B (b), solutions to bp of H_2O. Convenient system, shown in Figure 962.09C, consists of sheet Cu tank with 3 coils of 2/3 in. (10 mm) od Cu tubing, 12.5 ft. (3.8 m) long. Solder inlets and outlets where tubing passes through tank walls. Connect to reflux condenser and fill with H_2O. Keep H_2O boiling with two 750 watt thermostatically controlled hot plates. Use Tygon for inlet leads to reservoirs of H_2O, acid, and alkali; use gum rubber tubing for outlets. Capacity of preheater is adequate for 60 analyses in 8 h.

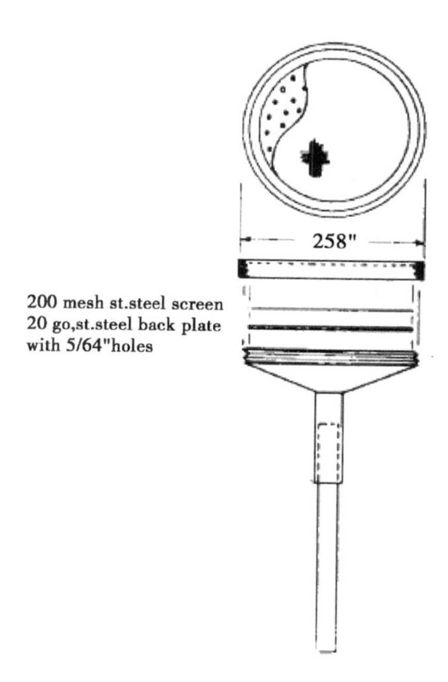

Figure 962.09A. Oklahoma State filter screen

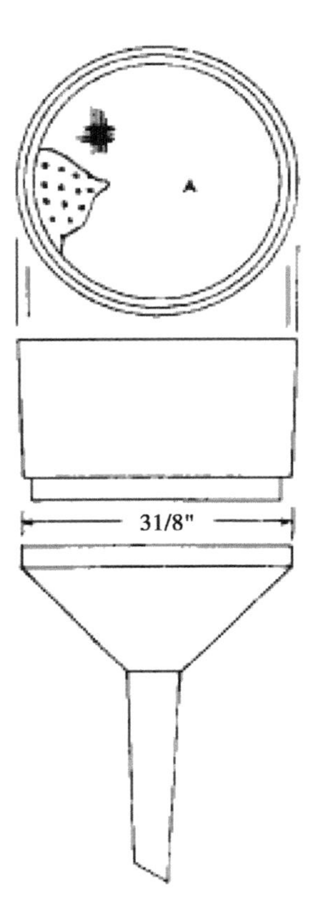

Figure 962.09B. Modified California State Buchner funnel, 2-piece polypropylene plastic, covered with 200-mesh screen, A, heat-sealed to edge of filtering surface

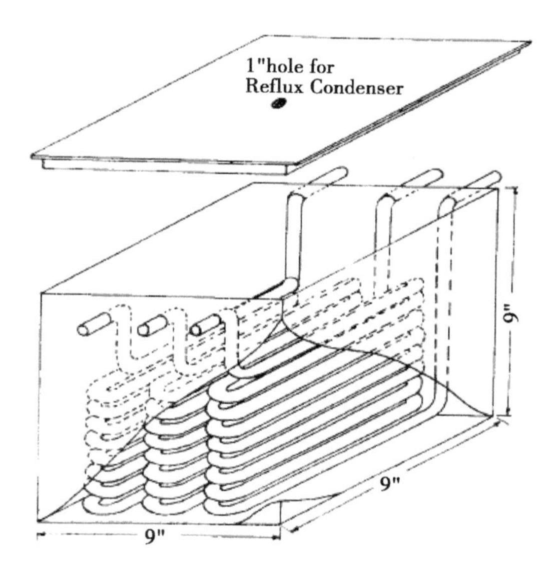

Figure 962.09C. Continuous heater for distilled water, 1.25% alkali, and 1.25% acid

Appendix 6 AOAC Official Method 2001.12 Determination of Water/Dry Matter (Moisture) in Animal Feed, Grain, and Forage (Plant Tissue) Karl Fischer Titration Methods First Action 2001

[The method is applicable to the determination of 1%~15% water in dry, ground animal feed and forage. It is not applicable for mineral mix feed. An estimated LOD is 0.25% and an estimated LOQ is 0.8% for 500 mg test portions (estimated LOD of 0.12% and estimated LOQ of 0.4% for 1g test portions).]

Caution: HYDRANAL©-Composite 5 contains 5 hazardous components—iodine, sulfur dioxide, imidazole, diethylene glycol monoethyl ether, and hydriodic acid, and should be handled with care.

See Tables 2001.12A-C for the results of the interlaboratory study supporting acceptance of the methods.

A. Principle

For Method I, water is extracted from the animal feed or forage material into methanol-formamide (1 + 1) directly in the Karl Fischer titration vessel by high-speed homogenization. The formamide content must not exceed 50% to maintain reaction conditions. A subsequent titration of the water is performed at 50℃ with one-component Karl Fischer reagent based on imidazole.

In Method II, boiling methanol is used instead of the methanol-formamide mixture at a higher temperature of 66℃ without high-speed homogenization. A totally tight titration vessel is critical.

These methods provide for the dry, ground test sample to be weighed and added directly to the titration cell to minimize contamination by atmospheric moisture. The reactions are:

$$CH_3OH + SO_2 + RN = [RNH]\ SO_3CH_3$$

where RN = imidazole base.

$$H_2O + I_2 + [RNH]\ SO_3CH_3 + 2RN$$
$$= [RNH]\ SO_4CH_3 + 2\ [RNH]\ I$$

Imidazole is used as the base to neutralize the acids and buffer the titration system. (*Note*: The pH of the system could be upset if a feed, such as a mineral mix, which was either a strong acid or base were to be analyzed.)

One source of interference is from aldehydes and ketones, which react with methanol to form acetals and ketals, respectively, and water. Do not use acetone to rinse glassware or equipment. The acetone residue reacts with methanol to form acetone dimethyl acetal and water, and thus bias water results high. Additional potential interferences include metal oxides, such as CaO, or other compounds that can form water when neutralized; or thiols, such as cysteine hydrochloride, or other compounds that can be oxidized by iodine. If interferences are suspected, verify by running the water standard in triplicate in the same solvent as used for the test sample immediately after the test sample, thereby retaining interfering compounds in the titration flask. In most cases, if interferences are present, they will affect the value for the water standard in the same manner as the test portion. Be alert to the possibility of potential interferences and run water standards frequently. Extracting replicates of a test sample in the same solvent may also magnify interference effects. If interfering compounds are found, additives or alternative solvents exist to deal with many.

Method I

Extraction into Methanol–Formamide (1 + 1)

Using High–Speed Homogenization

B. Apparatus

(a) *Karl Fischer titration–homogenization system.* —Metrohm 720 KFS Titrino, includes titration vessel LP with water jacket (50~150 mL), "snap–in" buret unit (10 mL), 703 titration stand with pump, cable with timer (Titrino – Polytron), or equivalent. *See* Figure 2001.12A (Brinkmann Instruments, Inc., One Cantiague Rd, PO Box 1019, Westbury, NY 11590–0207, www. brinkmann. com).

(b) *Circulating water bath.* —Maintaining 50℃ ± 1℃.

(c) *Homogenizer.* —Brinkmann Polytron homogenizer with PTA 20TSM foam–reducing generator with saw teeth and knives. Assembled with generator extending into the titration vessel, and adjusted to be ca 1 in. (2.5 cm) from the bottom of the titration vessel. Set to provide homogenization speed of 24 000 rpm. *See* Figure 2001.12A.

(d) *Glass weighing spoon.* —With opening for dispensing test portion into the titration flask through the septum stopper (Brinkmann), or equivalent.

(e) *Magnetic stirrer.*

(f) *Oven.* —103℃ ± 2℃.

C. Reagents

(a) *Karl Fischer reagent.* —One component, based on imidazole, with titer ca 5 mg H_2O/mL reagent, HYDRANAL© – Composite 5 (available from Riedel – de Haen 34805, Sigma Aldrich, St. Louis, MO), or equivalent.

(b) *Methanol.* —Anhydrous, for moisture determinations, water content not to exceed 0.05% (HYDRANAL–Methanol Dry; Riedel–de Haen 34741, Sigma Aldrich, St.

Louis, MO), or equivalent.

(c) *Formamide.*

(d) *Solvent.* —Methanol-formamide (1 + 1). Mix fresh each day.

(e) *Sodium tartrate dihydrate.* —Primary standard (water content, 15.66 ± 0.05%), HYDRANAL© - standard sodium tartrate - 2 - hydrate (Riedel - de Haen 34803, Aldrich Chemical Co., Milwaukee, WI), or equivalent.

(f) *Water standard.* —Water standard with certificate (water content, 10 mg/g), HYDRANAL©-water standard 10.0 (Riedel-de Haen 34849, Aldrich Chemical Co.), or equivalent.

D. Preparation of Test Sample

Grind test samples to pass a ≤1 mm opening. Mix well and transfer to a tightly sealed container.

E. Drying or Conditioning the Cell

Dispense sufficient methanol-formamide solvent into the titration vessel to immerse the homogenizer tip about 0.25 in. (8 mm). Close the cell to minimize the addition of atmospheric moisture. Heat to 50℃ ± 1℃. Dry the cell (including solvent, cell walls, electrode walls, generator, and cell atmosphere) by performing a complete run without test portion ("blank" run), including homogenization and titration as follows: Start the instrument and condition the solvent by titrating to remove moisture. As soon as instrument drift has stabilized, start the method. Homogenize at 24 000 rpm for 60 s. Enter "1" for test portion weight, and titrate the solvent again to remove any traces of water remaining. The end point is reached when no change in potential is observed for 10 s (titration system programmed for stop criterion: time; delay: 10 s). A dried titration cell has a maximum drift consumption of 5~10 μL Karl Fischer reagent per minute.

F. Standardization

Heat cell to 50℃±1℃. Dry the cell as in E. Depending on instrument, call up calibration mode. Condition solvent by titrating background moisture (press "start" key). Quickly weigh 150–250 mg of Na tartrate dihydrate standard into the glass weighing spoon and record weight of spoon and standard to the nearest 0.1 mg (S). Quickly transfer the weighed test portion into the titration flask through the septum stopper. Homogenize at 24 000 rpm for 60 s. Reweigh empty spoon to obtain tare weight (T) while homogenizing. Obtain the weight of standard material added by subtracting tare weight (T) from weight of spoon plus standard (S). Record weight of standard material (S – T) in mg to the nearest 0.1 mg. Stop the homogenizer after 60 s. Enter weight into instrument, turn on the stirrer and start the titration. Before reaching the endpoint of the titration, homogenize momentarily to rinse down moisture that may have been under the cap. Titrate to same endpoint as in E, recording volume of reagent required for the titration (mL reagent) in mL to the nearest 0.001 mL. Repeat 4 times. Calculate titer, then average the 5 values. The relative standard deviation should be <2%.

$$\text{Titer} = \frac{\text{mgH}_2\text{O}}{\text{mL reagent}} = \frac{\text{mgNa}_2\text{C}_4\text{H}_4\text{O}_6 \cdot 2\text{H}_2\text{O} \times 0.1566}{\text{mL reagent}}$$

Where mg $Na_2C_4H_4O_6 \cdot 2H_2O$ is S – T, in mg.

G. System Suitability

Heat cell to 50℃±1℃. Dry the cell as in E. Check drift in the titration cell. A dried titration cell should have a maximum drift consumption of 5~10 μL Karl Fischer reagent/ min. Analyze a water standard as follows: Immediately after drying the cell, break open the standard ampoule at the white ring and take 1~2 mL of standard with syringe which has been predried in a 103℃ oven. Rinse the syringe and discard the standard solution. Draw the remaining water standard (~6 mL) into the syringe and weigh accurately by placing the syringe into a beaker on the balance pan (S_S). Quickly add ~2 mL water standard through the septum keeping the tip of the syringe below the surface of the solvent. Carefully withdraw the syringe tip, reweigh the syringe and record the weight (S_1). Obtain the weight of standard solution ($S_s - S_1$) by subtracting the weight (S_1) from the weight of the syringe plus standard (S_S). Record the water standard weight to the nearest 0.1 mg. Enter the weight into the instrument, start the stirrer, and begin the titration. Record the volume of titrant (V_1). Carry out the titration procedure two additional times, recording weights of the syringe after each subsequent addition (S_2, S_3) and the respective volume of titrant (V_2, V_3). Calculate the percent recovery as follows:

$$\text{Water standard, g} = S_S - S_1,\ S_1 - S_2,\ S_2 - S_3$$

$$\text{Rec., \%} = \frac{\dfrac{V_a \times \text{titre}}{\text{g standard}}}{\text{oertified value, mg/s}} \times 100 = \frac{\text{mg/gg } H_2O \text{ found}}{\text{oertified value, mg/s}} \times 100$$

Average % recovery should be 100% ± 1%. If the system is not within specifications, correct before continuing with determinations. If the % recovery on the water standard is within specification, it is not necessary to perform a blank run (with no material), since the water standard indicates the condition of the system and running a blank will provide no additional information.

H. Determination

Dry the cell as in E. Depending on the instrument call up the sample analysis mode. Quickly weigh ~ 0.5 g test portion (to contain ~25 to 50 mg water) into the glass weighing spoon and record weight of the spoon plus the test portion (W). Quickly add weighed test portion into the titration flask through the septum stopper. Homogenize at 24 000 rpm for 60 s. Reweigh empty spoon and record tare weight (T) while homogenizing. Obtain the test portion weight by subtracting tare weight (T) from weight of spoon plus test portion (W). Record weight (W − T) in g to the nearest 0.1 mg. Stop homogenizer after 60 s. Enter weight into the instrument and start the titration. Before reaching the end point of the titration, homogenize momentarily to rinse down moisture that may have been under the cap. This ensures that the moisture of the particles under the cap or hanging on the glass walls above the liquid surface is titrated. The end point is reached when no change in potential is observed for 10 s (stop criterion: time; delay: 10 s). Record the volume of titrant (V). Repeat determination in triplicate. The relative standard deviation of replicates should be <5%. The cell need not be emptied between each titration. Usually about 3 titrations can be performed before the cell requires emptying and replenishing.

I. Calculations

$$\text{mg } H_2O = V \times \text{titer}$$

$$\% \ H_2O = \frac{V \times \text{titer}}{10 \times \text{test potlion wt}}$$

$$\text{Dry matter, } \% = 100 - \% \ H_2O$$

Where V is the volume of titrant in mL and test portion weight is W − T, in g.

Method II

Boiling Methanol Extraction Alternative

First Action 2002

J. Apparatus

(a) *Karl Fischer titration system.* —Metrohm 720 KFS Titrino "snap-in" burette unit (10 mL), 703 titration stand, or equivalent. Incorporate a titration vessel with water jacket (50–150 mL) and condenser (Horst Kuhn, Am Tutberg 32, 3032 Fallingbostel, Germany) or a 4 neck flask with a heating mantle. *See* Figure 2001.12B.

(b) *Circulating water bath.* —Maintaining 75℃ ± 1℃.

(c) *Glass weighing spoon.* —With opening for dispensing test portion into the titration flask through the septum stopper (Brinkmann 6.2412.000 or equivalent).

(d) *Magnetic stirrer.*

K. Reagents

See C.

L. Preparation of Test Sample

See D.

M. Drying or Conditioning the Cell

Dispense ~50 mL methanol into the titration vessel. Close the cell to minimize the addition of atmospheric moisture. Heat until the MeOH begins to boil. Dry the cell (including solvent, cell walls, electrode walls, generator, and cell atmosphere) by titrating to dryness. The end point is reached when no change in potential is observed for 10 s (titration system programmed for" stop criterion: time; delay: 10 s). A dried titration cell has a maximum drift consumption of 5~10 μL Karl Fischer reagent per minute.

N. Standardization

Heat cell to 66℃ ± 1℃ (boiling point of methanol). Dry the cell as described in M. Depending on instrument, call up calibration mode. Condition solvent by titrating background moisture (hit "start"). Switch off the heating system and when the methanol stops boiling, quickly weigh 150~250 mg of sodium tartrate dihydrate standard into the glass weighing spoon and record weight of spoon and standard to the nearest 0.1 mg

(S). Quickly transfer the weighed test portion into the titration flask through the septum stopper. Reweigh empty spoon to obtain tare weight (T) and obtain the weight of standard material added by subtracting tare weight (T) from weight of spoon plus standard (S). Record weight of standard material (S − T) in mg to the nearest 0.1 mg. Enter weight into instrument, start the stirrer and begin the titration. Titrate to same endpoint as described in M, recording volume of reagent required for the titration (mL reagent) in mL to the nearest 0.001 mL. Repeat 4 times. Calculate titer as in F, then average the 5 values. The relative standard deviation should be <2%.

O. System Suitability

Heat cell to 66℃ ± 1℃ (boiling point of methanol). Dry the cell as described in M. Check drift in the titration cell. A dried titration cell should have a maximum drift consumption of 5 ~ 10 μL Karl Fischer reagent/min. Analyze a water standard as follows: Immediately after drying the cell, switch off the heating system and after the methanol stops an active boil, break open the standard ampoule at the white ring and take 1~2 mL of standard with syringe which has been predried in a 103℃ oven. Rinse the syringe and discard the standard solution. Draw the remaining water standard (~6 mL) into the syringe and weigh accurately by placing the syringe into a beaker on the balance pan (S_S). Quickly add ~2 mL water standard through the septum keeping the tip of the syringe below the surface of the solvent. Carefully withdraw the syringe tip, reweigh the syringe and record the weight (S_1). Obtain the weight of standard solution ($S_s - S_1$) by subtracting the weight (S_1) from the weight of the syringe plus test portion (S_S). Record the water standard weight to the nearest 0.1 mg, and enter the weight into instrument. Turn on the heating system, start the stirrer and begin the titration as soon as the methanol returns to an active boil. Record the volume of titrant (V_1). Carry out the titration procedure 2 additional times, recording weights of the syringe after each subsequent addition (S_2, S_3) and the respective volume of titrant (V_2, V_3). Calculate the percent recovery as in G.

Average % recovery should be 100% ± 1%. If system is not within specifications, correct before continuing with determinations. If the % recovery on the water standard is within specification, it is not necessary to perform a blank run (with no material), since the water standard indicates the condition of the system and running a blank will provide no additional information.

P. Determination

Dry the cell as described in M. Depending on the instrument call up the sample analysis mode. After switching off the heating system and the methanol stops an active boil, quickly weigh ~0.5 g test portion (to contain ~25 to 50 mg water) into the glass weighing spoon and record weight of the spoon plus the test portion (W). Quickly add weighed test portion into the titration flask through the septum stopper. Reweigh empty spoon and record tare weight (T). Obtain the test portion weight by subtracting tare weight (T) from weight of spoon plus test portion (W). Record weight (W − T) in g to the nearest 0.1 mg. Enter weight into the instrument, start the stirrer, turn on the heating system, and begin the titration as soon as the methanol returns to an active boil. The end point is reached when no change in potential is observed for 10 s (stop criterion: time; delay: 10 s). Record the volume of titrant (V). Repeat determination in triplicate. The relative standard deviation of replicates should be <5%. The cell need not be emptied between each titration. Usually about 3 titrations can be performed before the cell requires emptying and replenishing.

Q. Calculations

See I.

Reference:

J. AOAC Int. (future issue).

Table 2001.12A: Statistical results for interlaboratory study on determination of water in animal feed, cereal, and forage (blind duplicate design)

ID	Mean	No. labs[a(b)]	s_r	RSD_r, %	s_R	RSD_R, %	HORRAT
Soybeans	6.99	9	0.10	1.38	0.44	6.31	2.11
Urea feed	4.58	9	0.19	4.11	0.49	10.73	3.37
Grass hay	6.35	9	0.18	2.89	0.48	7.59	2.51
Cereal grains	8.58	9	0.39	4.49	0.46	5.35	1.85
Alfalfa silage	7.26	7 (2)	0.14	1.98	0.72	9.85	3.32
Corn silage	6.40	9	0.22	3.40	0.47	7.34	2.43
Cat food	6.90	9	0.24	3.50	0.43	6.27	2.10
Milk replacer	5.21	9	0.32	6.13	0.42	7.99	2.56
Alfalfa hay	7.36	9	0.51	6.99	0.66	8.95	3.02
Oats	7.53	8 (1)	0.09	1.14	0.61	8.12	2.75

[a(b)] a = Number of laboratories retained after eliminating outliers, (b) = number of laboratories removed as outliers.

Table 2001.12B: Statistical results for collaborative study on determination of dry matter in animal feed, cereal, and forage (blind duplicate design)

Feeds	Mean	Lab[a]	s_r	RSD_r, %	s_R	RSD_R, %	HORRAT
Soybeans	93.01	9	0.10	0.10	0.44	0.47	0.23
Urea feed	95.42	9	0.19	0.20	0.49	0.52	0.26
Grass hay	93.65	9	0.18	0.20	0.48	0.51	0.25
Cereal grains	91.42	9	0.39	0.42	0.46	0.50	0.25
Alfalfa silage	92.74	7 (2)	0.14	0.16	0.72	0.77	0.38
Corn silage	93.60	9	0.22	0.23	0.47	0.50	0.25
Cat food	93.10	9	0.24	0.26	0.43	0.46	0.23
Milk replacer	94.79	9	0.32	0.34	0.42	0.44	0.22
Alfalfa hay	92.64	9	0.51	0.56	0.66	0.71	0.35
Oats	92.47	8 (1)	0.09	0.09	0.61	0.66	0.33

[a] Number of laboratories retained after eliminating outliers. Number of laboratories removed as outliers shown in parentheses.

Table 2001.12C: Statistical results for interlaboratory study on determination of water in animal feed, cereal, and forage, including data from laboratory using alternative extraction (blind duplicate design)

ID	Mean	No. labs[a(b)]	sr	RSDr, %	s_R	RSD_R, %	HORRAT
Soybeans	7.01	10	0.10	1.41	0.42	6.03	2.02
Urea feed	4.57	10	0.18	3.97	0.47	10.20	3.20
Grass hay	6.34	10	0.28	4.44	0.48	7.59	2.51
Cereal grains	8.60	10	0.37	4.26	0.44	5.08	1.76
Alfalfa silage	7.40	9 (1)	0.15	2.05	0.68	9.17	3.10
Corn silage	6.43	10	0.21	3.23	0.45	7.02	2.32
Cat food	6.94	10	0.23	3.33	0.44	6.27	2.10
Milk replacer	5.18	10	0.30	5.85	0.40	7.76	2.49
Alfalfa hay	7.40	10	0.49	6.64	0.64	8.59	2.90
Oats	7.51	10	0.15	1.99	0.57	7.54	2.55

[a(b)] a = Number of laboratories retained after eliminating outliers, (b) = number of laboratories removed as outliers.